Survivability under Overheating

Survivability under Overheating—The impact of Regional and Global Climate Change on Vulnerable and Low Income Population

Editors

Afroditi Synnefa
Shamila Haddad
Priya Rajagopalan
Matthaios Santamouris

MDPI • Basel • Beijing • Wuhan • Barcelona • Belgrade • Manchester • Tokyo • Cluj • Tianjin

Editors
Afroditi Synnefa
University of New South Wales
Australia

Shamila Haddad
University of New South Wales
Australia

Priya Rajagopalan
RMIT University
Australia

Matthaios Santamouris
University of New South Wales
Australia

Editorial Office
MDPI
St. Alban-Anlage 66
4052 Basel, Switzerland

This is a reprint of articles from the Special Issue published online in the open access journal *Climate* (ISSN 2225-1154) (available at: https://www.mdpi.com/journal/climate/special_issues/survivability).

For citation purposes, cite each article independently as indicated on the article page online and as indicated below:

LastName, A.A.; LastName, B.B.; LastName, C.C. Article Title. *Journal Name* **Year**, *Volume Number*, Page Range.

ISBN 978-3-03943-869-3 (Hbk)
ISBN 978-3-03943-870-9 (PDF)

Contents

About the Editors . . . . **vii**

Afroditi Synnefa, Shamila Haddad, Priyadarsini Rajagopalan and Mattheos Santamouris
SI: Survivability under Overheating: The Impact of Regional and Global Climate Change on the Vulnerable and Low-Income Population
Reprinted from: *Climate* **2020**, *8*, 122, doi:10.3390/cli8110122 . . . . **1**

Afifa Mohammed, Gloria Pignatta, Evangelia Topriska and Mattheos Santamouris
Canopy Urban Heat Island and Its Association with Climate Conditions in Dubai, UAE
Reprinted from: *Climate* **2020**, *8*, 81, doi:10.3390/cli8060081 . . . . **5**

Kai Gao, Mattheos Santamouris and Jie Feng
On the Efficiency of Using Transpiration Cooling to Mitigate Urban Heat
Reprinted from: *Climate* **2020**, *8*, 69, doi:10.3390/cli8060069 . . . . **33**

Andri Pyrgou and Mattheos Santamouris
Probability Risk of Heat- and Cold-Related Mortality to Temperature, Gender, and Age Using GAM Regression Analysis
Reprinted from: *Climate* **2020**, *8*, 40, doi:10.3390/cli8030040 . . . . **49**

Jeremy S. Hoffman, Vivek Shandas and Nicholas Pendleton
The Effects of Historical Housing Policies on Resident Exposure to Intra-Urban Heat: A Study of 108 US Urban Areas
Reprinted from: *Climate* **2020**, *8*, 12, doi:10.3390/cli8010012 . . . . **59**

Andri Pyrgou, Mattheos Santamouris, Iro Livada and Constantinos Cartalis
Retrospective Analysis of Summer Temperature Anomalies with the Use of Precipitation and Evapotranspiration Rates
Reprinted from: *Climate* **2019**, *7*, 104, doi:10.3390/cli7090104 . . . . **75**

About the Editors

Afroditi Synnefa is a physicist specializing in energy and environmental issues with focus on the built environment. She has a Ph.D. in Physical Sciences from the National and Kapodistrian University of Athens, Greece. She is currently Adjunct Senior Lecturer in the High-Performance Architecture Research Cluster at the Faculty of the Built Environment, at UNSW, Sydney, Australia. Her main research interests include development of innovative materials for the built environment to mitigate the urban heat island effect, optical and thermal characteristics of materials for the built environment, urban climate change and heat mitigation technologies, energy efficient building, indoor environmental quality, and sustainable energy technologies and energy saving technologies in buildings and settlements. She has contributed to more than 20 research projects funded by national, European, and international organizations and companies. She is the Technical Committee leader of the European Cool Roofs Council, where she also acts as a consultant. She has teaching experience at undergraduate and postgraduate levels and in professional education and e-learning. She has published a large number of papers in scientific journals and as international conference proceedings.

Shamila Haddad holds a Ph.D. in Architectural Sciences from the University of New South Wales (UNSW), where she currently serves as Associate Lecturer, focusing on High Performance Architecture at the Faculty of Built Environment, UNSW. Her work aims to plan, design, and manage sustainable buildings and cities. Her research and teaching expertise and interests are related to environmental sustainability, building performance, climate change, urban heat mitigation technologies, indoor environmental and air quality, thermal comfort, and healthy, energy efficient, and innovative buildings. Dr Haddad's publication track record features papers on the indoor air and environmental quality research fields, urban climate, urban microclimate simulations, and developing advanced heat mitigation and adaptation technologies, published in scientific journals and international conference proceedings.

Priya Rajagopalan is a building scientist with extensive experience in energy and indoor environmental quality of buildings, urban climatology, and urban thermal balance. She is a Professor and the Director of the Sustainable Building Innovation Lab (SBi Lab) at the School of Property, Construction and Project Management, RMIT University, Melbourne, Australia. Her research interests span from sustainable building design, energy benchmarking and labeling, urban heat islands, urban greenery, building performance simulation, lighting, and acoustics to citizen science in urban microclimate monitoring. She has coordinated several projects in the area of energy benchmarking and labeling of buildings, indoor air quality as well as urban heat island in the tropical and temperate climates. She has published more than 90 scientific papers. Priya is also the Vice President of the Architectural Science Association (ANZAScA).

Matthaios Santamouris is a Scientia Professor of High Performance Architecture at UNSW, and past Professor at the University of Athens, Greece. He has served as Visiting Professor of Cyprus Institute, Metropolitan University London, Tokyo Polytechnic University, Bolzano University, Brunnel University, Seoul University, and National University of Singapore. He is Past President of the National Center of Renewable and Energy Savings of Greece. He serves as Editor in Chief of Energy and Buildings, Associate Editor of Solar Energy, and Member of the Editorial Boards of 14 journals and was a former Editor in Chief of Advances in Building Energy Research. He is Editor of a series of books on Buildings published by Earthscan Science Publishers as well as editor and author of 14 international books published by Elsevier, Earthscan, and Springer. He has authored 342 scientific articles published in journals and is reviewer of research projects in 29 countries, including USA, UK, France, Germany, Canada, and Sweden.

Editorial

SI: Survivability under Overheating: The Impact of Regional and Global Climate Change on the Vulnerable and Low-Income Population

Afroditi Synnefa [1], Shamila Haddad [1,*], Priyadarsini Rajagopalan [2] and Mattheos Santamouris [1]

[1] Faculty of the Built Environment, University of New South Wales, Sydney, NSW 2052, Australia; a.synnefa@unsw.edu.au (A.S.) m.santamouris@unsw.edu.au(M.S.)
[2] Sustainable Building Innovation Lab, School of Property, Construction and Project Management, RMIT University, 124 La Trobe Street, Melbourne, VIC 3000, Australia; priyadarsini.rajagopalan@rmit.edu.au
* Correspondence: s.haddad@unsw.edu.au

Received: 19 October 2020; Accepted: 21 October 2020; Published: 24 October 2020

Abstract: The present special issue discusses three significant challenges of the built environment, namely regional and global climate change, vulnerability, and survivability under the changing climate. Synergies between local climate change, energy consumption of buildings and energy poverty, and health risks highlight the necessity to develop mitigation strategies to counterbalance overheating impacts. The studies presented here assess the underlying issues related to urban overheating. Further, the impacts of temperature extremes on the low-income population and increased morbidity and mortality have been discussed. The increasing intensity, duration, and frequency of heatwaves due to human-caused climate change is shown to affect underserved populations. Thus, housing policies on resident exposure to intra-urban heat have been assessed. Finally, opportunities to mitigate urban overheating have been proposed and discussed.

Keywords: climate change; urban heat island; mitigation; resilience; survivability; low-income population

---

1. Introduction

The so-called urban heat island (UHI) phenomenon has been known for decades, and many studies have been developed worldwide to address urban overheating. However, rapid urbanization, combined with a continuous increase in the anthropogenic heat in cities, has intensified the phenomenon's magnitude and aggravated its impact on energy, environment, comfort, and health. Further, global climate change has caused a significant increase in the frequency, magnitude, and duration of extreme heat events intensifies UHI's magnitude, especially during heat waves because of the critical synergetic effects.

An increase in the ambient temperature and more frequent heat waves significantly impact the energy consumption and environmental quality of buildings and increase the vulnerability of the local population [1]. Energy poverty refers to the conditions where households cannot afford their basic energy needs. In other words, it refers to the "inability to adequately meet household energy needs" [2]. Research has shown a strong correlation between energy poverty and building performance and the local climate change [1]. Houses with high energy consumption are a severe burden to low-income populations who can hardly afford to satisfy energy needs for proper indoor temperature and environmental conditions. In parallel, low-income populations mainly live in quite deprived and degraded urban areas, where the phenomenon of the urban heat island is quite strong. Exposure to higher summer outdoor temperatures considerably increases the low-income population's vulnerability and puts their health conditions under stress, leading to increased mortality rates.

Heat mitigation and adaptation technologies to upgrade the environmental and climatic conditions in deprived urban zones and improve the energy performance and indoor environmental quality of low-income households are essential technological responses to the problem. This special issue aims to discuss and present the problems and highlight the need to develop heat mitigation plans to counterbalance the impacts of overheating.

2. Special Issue Content

Five papers have been published in this special issue discussing several topics related to climate change, UHI, heat mitigation technology, survivability, resilience, and low-income population. Climate change and urbanization affect the thermal-energy balance of the built environment. This is a primary environmental concern, as it has negative impacts on energy, environment, comfort, and health. The UHI can raise the temperature in cities, which is a significant problem worldwide. In the first paper by Mohammed et al. [3], the UHI magnitude and its association with the main meteorological parameters (i.e., temperature, wind speed, and wind direction) in the hot arid climate of the United Arab Emirates have been discussed. Several studies have been developed to examine the impacts of different urban heat mitigation technologies and evaluate their effectiveness. Among all, trees are useful for the mitigation of urban overheating via transpiration and shading. The paper by Gao et al. [4] explores the transpiration cooling of large trees in urban environments where the sea breeze dominates the climate. This study highlights the importance of considering synoptic conditions when planting trees for mitigation purposes.

Global climate change increases the frequency, magnitude, and duration of extreme heat events. Low-income populations typically live in degraded urban areas, where the phenomenon of the urban overheating is strong. Higher summer outdoor temperatures increase the vulnerability of the low-income population and adversely affect their health.

Longer and more frequent heat waves due to human-caused climate change put historically underserved populations in a heightened state of precarity, as studies observe that vulnerable communities are disproportionately exposed to extreme heat. The paper by Hoffman et al. [5] explores how historic policies of redlining help to explain current patterns of intra-urban heat and the extent to which these patterns were consistent across 108 US urban areas. It reveals that historical housing policies may be directly responsible for disproportionate exposure to current heat events.

The increased frequency of temperature extremes is concerning as they are associated with increased morbidity and mortality. The relationship between hot and cold conditions and mortality from respiratory and cardiovascular causes is well established. Pyrgou and Santamouris [6] examined the heat and cold-related mortality risk of different age groups subject to cold and heat extremes and compared them between the two genders.

Further to the above, drought and extreme temperature forecasting is essential for water management and the prevention of health risks, especially in a period of observed climatic change. A large precipitation deficit and increased evapotranspiration rates in the preceding days contribute to exceptionally high temperature anomalies in the summer, above the average local maximum temperature for each month. The study by Pyrgou et al. [7] investigated droughts and extreme temperatures in Cyprus and suggested a different approach in determining the lag period of summer temperature anomalies and precipitation.

3. Conclusions

This special issue aims to highlight the issues related to local climate change and extreme heat, to enhance survivability against the effects of insecurities in climate change and energy poverty. The papers presented here cover different climatic contexts and discuss several issues related to changing climate. This knowledge is vital to develop appropriate strategies for managing and preventing health risks caused by extreme temperatures. The substantial impact that energy poverty can have on population health highlights the need to adopt policies and practices that protect people from energy insecurity and climate change. Many countries have recognized the importance of energy-efficient techniques, adaptation, and mitigation technologies to drastically improve energy poverty. Therefore, the use of energy-efficient technology will help to reduce vulnerability to the changing climate and revitalize the low-income households' economic situation.

Author Contributions: Conceptualization, M.S.; writing—original draft preparation, A.S., S.H., P.R., and M.S.; writing—review and editing, A.S., S.H., P.R., and M.S. All authors have read and agreed to the published version of the manuscript.

Funding: This research received no external funding.

Acknowledgments: Authors thank editors for reviewing submitted manuscripts.

Conflicts of Interest: The authors declare no conflict of interest.

References

1. Santamouris, M. Innovating to zero the building sector in Europe: Minimising the energy consumption, eradication of the energy poverty and mitigating the local climate change. *Sol. Energy* **2016**, *128*, 61–94. [CrossRef]
2. Hernández, D. Understanding 'energy insecurity' and why it matters to health. *Soc. Sci. Med.* **2016**, *167*, 1–10. [CrossRef] [PubMed]
3. Mohammed, A.; Pignatta, G.; Topriska, E.; Santamouris, M. Canopy Urban Heat Island and Its Association with Climate Conditions in Dubai, UAE. *Climate* **2020**, *8*, 81. [CrossRef]
4. Gao, K.; Santamouris, M.; Feng, J. On the Efficiency of Using Transpiration Cooling to Mitigate Urban Heat. *Climate* **2020**, *8*, 69. [CrossRef]
5. Hoffman, J.; Shandas, V.; Pendleton, N. The effects of historical housing policies on resident exposure to intra-urban heat: A study of 108 US urban areas. *Climate* **2020**, *8*, 12. [CrossRef]
6. Pyrgou, A.; Santamouris, M. Probability Risk of Heat-and Cold-Related Mortality to Temperature, Gender, and Age Using GAM Regression Analysis. *Climate* **2020**, *8*, 40. [CrossRef]
7. Pyrgou, A.; Santamouris, M.; Livada, I.; Cartalis, C. Retrospective Analysis of Summer Temperature Anomalies with the Use of Precipitation and Evapotranspiration Rates. *Climate* **2019**, *7*, 104. [CrossRef]

Article

Canopy Urban Heat Island and Its Association with Climate Conditions in Dubai, UAE

Afifa Mohammed [1,*], Gloria Pignatta [1], Evangelia Topriska [2] and Mattheos Santamouris [1]

[1] Faculty of Built Environment, University of New South Wales (UNSW), Sydney, NSW 2052, Australia; g.pignatta@unsw.edu.au (G.P.); m.santamouris@unsw.edu.au (M.S.)

[2] Department of Architectural Engineering, Faculty of Energy, Geoscience, Infrastructure and Society, Heriot-Watt University, Dubai International Academic City, Dubai 294345, UAE; e.topriska@hw.ac.uk

* Correspondence: a.mohammed@student.unsw.edu.au

Received: 25 May 2020; Accepted: 24 June 2020; Published: 26 June 2020

Abstract: The impact that climate change and urbanization are having on the thermal-energy balance of the built environment is a major environmental concern today. Urban heat island (UHI) is another phenomenon that can raise the temperature in cities. This study aims to examine the UHI magnitude and its association with the main meteorological parameters (i.e., temperature, wind speed, and wind direction) in Dubai, United Arab Emirates. Five years of hourly weather data (2014–2018) obtained from weather stations located in an urban, suburban, and rural area, were post-processed by means of a clustering technique. Six clusters characterized by different ranges of wind directions were analyzed. The analysis reveals that UHI is affected by the synoptic weather conditions (i.e., sea breeze and hot air coming from the desert) and is larger at night. In the urban area, air temperature and night-time UHI intensity, averaged on the five year period, are 1.3 °C and 3.3 °C higher with respect to the rural area, respectively, and the UHI and air temperature are independent of each other only when the wind comes from the desert. A negative and inverse correlation was found between the UHI and wind speed for all the wind directions, except for the northern wind where no correlation was observed. In the suburban area, the UHI and both temperatures and wind speed ranged between the strong and a weak negative correlation considering all the wind directions, while a strong negative correlation was observed in the rural area. This paper concludes that UHI intensity is strongly associated with local climatic parameters and to the changes in wind direction.

Keywords: subtropical desert climate; urban overheating; cluster analysis; air temperature; wind speed and wind directions; synoptic conditions

---

1. Introduction

As observed in many cities globally, rapid urbanization has produced negative effects on the climate and the local microclimate. Currently, the urban population exceeds 50% of the total of the world's population and by 2050 it is expected to rise above 60%. This means that with urban development worldwide and at the current rate of population growth, another 2.5 billion people will be living in urban areas by 2050 [1]. Consequently, this urban expansion will exacerbate the hostile impact that human activities are already having on environmental systems.

Rapid urbanization has boosted regional climate change. In particular, the increase in urban heat island (UHI) intensity is strictly connected with the urbanization growth given the derived increase in anthropogenic heat emissions, land-use change (with the associated decrease in the vegetation and the albedo of the built area), and changes in the advection.

The UHI phenomenon has been documented in more than 400 cities around the world [2]. Its impact is closely related to land cover which controls the energy budget on the earth's surface. The surface energy budget difference between the urban and rural zone, caused by various thermal-optical surface characteristics, leads to the occurrence of the UHI phenomenon [3]. Thus, urban areas are hotter than their undeveloped surroundings. This phenomenon has a significant impact not only on the environmental quality of cities, but also on energy, thermal comfort, and health [4–7]. In particular, the UHI increases the demand for peak-time electricity and the consumption of cooling energy in buildings, intensifies the concentration of various harmful pollutants, increases the ecologically harmful footprint of cities, and has a significant impact on health [8]. Most cities are sources of pollution and heat, released from buildings and roads. The manifestation of the UHI phenomenon is influenced by several factors including climate variables (i.e., air temperature, air relative humidity, wind speed) and the related local synoptic weather conditions [9,10], thermal-optical characteristics of the materials, the magnitude of the anthropogenic heat released, and the existing heat sources in the areas [2].

On the other hand, the construction and building industries are indispensable to everyday life and are essential to future social and technological developments. However, they are responsible for a great deal of pollution, generation of waste, and the consumption of energy and natural resources accounting for 30%–40% of the overall consumption on the planet [11,12]. Moreover, these industries have negative effects on the local and global climate [12]. The International Energy Agency estimates that the growth in energy consumption will be around 38.4 PWh in urbanized countries by 2040, while energy consumption globally was about 23.7 PWh in 2010 [13]. These industries account for 38% of all greenhouses gas emissions. They also contribute significantly to increase the temperature of cities by producing the urban heat island (UHI) phenomenon [8].

Population growth and rural depopulation will lead to greater energy consumption, particularly in hot and dry cities which require more cooling loads; in addition, this increases the number of gas emissions and pollution that negatively interact with the phenomenon of the UHI [14]. Furthermore, the expansion of urban areas will require a more complex energy infrastructure to meet demand [12,15].

Recently, many studies have demonstrated that per degree of temperature increase, the peak electricity demand increases by between 0.45% and 4.6% due to the UHI [16]. Moreover, it is expected that by 2050, there will be a massive increase of about 750% and 275% globally in the cooling requirements of the residential and commercial sectors, respectively [17]. Furthermore, the UHI has a significant impact on human health, as several studies have shown that high temperatures lead to a great increase in the number of people suffering from heat exhaustion and heatstroke in urban areas [18]. Moreover, it has been pointed out that there is a strong correlation between the increase in human mortality and the rise in urban temperatures [19]. Several studies in various Asian cities have shown that temperatures above 29 °C can increase the mortality rate between 4.1% and 7.5% per 1 °C increase in temperature [20]. Hence, from the aforementioned discussion, it is clear that cities need specific controls and strategies to mitigate the negative impacts of the UHI phenomenon [12].

Among the performed studies on UHI, three different definitions of UHI are usually considered, such as the boundary UHI for mesoscale analysis [21], the canopy UHI for microscale analysis [22], and the surface UHI [23]. Different methods have been reported [24] to identify the different UHI types, including direct and indirect methods, numerical modeling, and estimates based on empirical models.

In contrast with the numerous studies on UHI performed in temperate regions, only a few studies have been concentrated in the observation of the UHI intensity in desert regions [25,26]. Some of those studies state that the UHI presents a diurnal and seasonal cycle in the Gulf area. The canopy level of the UHI of Muscat, Oman, reaches the peak approximately 7 h after sunset based on summer meteorological observations. Muscat city is characterized by low ventilation, many business activities, multi-storied buildings, heavy traffic, and topography factors [27]. In another study in Bahrain, numerical modeling was performed and the results show an increase in the simulated average air temperature of about 2 °C–5 °C when assessing the impact of the urbanization and that the canopy UHI magnitude is enhanced by various urban activities such as construction processes, vegetation shrinkage, and sea reclamation [28].

Given the lack of information about UHI intensity concepts in the Middle East in general [28] and specifically in Dubai, United Arab Emirates (UAE), this paper aims to quantify and analyze the intensity of the canopy UHI in Dubai, highly influenced by the urban geometry and physical properties of the built environment, and its relation with the main climatic parameters continuously monitored for a period of five years.

2. Materials and Methods

2.1. Geographical Location, Population, and Climate of Dubai

Dubai, one of the seven emirates and the second largest city of the United Arab Emirates (UAE), has an area of 4114 km^2, accounting for 5% of the overall area of the country. Dubai city is recognized as the economic capital of the UAE. Situated on the Tropic of Cancer, between 25°16′ N and 55°18′ E, it has a coastal length of 72 km on the eastern coast of the Arabian Peninsula and faces the South-West of the Arabian Gulf [29,30].

According to the census conducted by the Statistics Centre of Dubai, Dubai is the city with the highest population density in the UAE. The population of Dubai increased by 27% from 2,327,350 to 3,192,275 between 2014 and 2018 inclusively [31].

The Dubai region has a subtropical desert climate, with hot and humid summers and warm winters [30]. The air temperature ranges between 10 °C to 30 °C in the winter season and increases up to 48 °C in the summer season [29]. The hot period starts with average daily temperatures over 37 °C from 18 May to 23 September, lasting for 4.1 months; the cool season starts with average temperatures of less than 27 °C from December 4th to March 8th and lasts for 3.1 months [32]. Winter is characterized by rainfall and fog, while in summer, the relative humidity reaches 80%–90% [29]. The average rainfall ranges between 13 and 17 mm [32].

2.2. Meteorological Stations and Data Analysis

In this study, climate data (i.e., air temperature, relative humidity, wind speed, and wind direction) were collected by three meteorological stations in Dubai (i.e., Dubai International Airport station, Al-Maktoum International Airport station, and Saih Al-Salem Station) belonging to the UAE National Center of Meteorology (NCM). Hourly data of each microclimate parameter were collected from each meteorological station and analyzed in the MS Excel tool for a period of five years from 2014 to 2018 (i.e., 43,824 values recorded in total for each climate variable). Data of cloud coverage were also collected by the two stations of Dubai Airport and Al-Maktoum Airport, while the solar radiation that equally affects the three areas of Dubai was not measured by any meteorological station. In this study, no further tools were used for the data analysis except for MS Excel.

Table 1 summarizes the geographical information of the three meteorological stations [33], while Figure 1 shows their locations in Dubai. All three stations are installed at a height of about 10 m from the ground, allowing to perform the microscale analysis of the UHI phenomenon, investigating the canopy UHI and its association with the climate parameters measured at the same height.

Station 1 (Table 1) is located within Dubai International Airport, one of the largest and most modern airports in the world, located in Al Garhoud City, which is a commercial and residential region of Dubai, approximately 5 km (2.9 miles) East of the Dubai's CBD (central business district), and around 19 m above sea level [34]. A total of 15,000 solar panels were installed on the roof of the airport building 2 km away from the meteorological station [35].

Station 2 (Table 1) is located within the Al Maktoum International Airport, 37 km (23 miles) South-West of Dubai. The airport is surrounded by a mixed-use area, made of residential and commercial buildings; it is the major part of the Dubai World Central in the Jebel Ali zone with a total area of around 14,000 hectares (35,000 acres). This area could be described as an open area. It is almost at a 16 km (10 miles) distance from the sea and has no surrounding buildings or major obstacles that could interfere with the wind flowing from different directions [36].

Station 3 (Table 1) is located within Saih Al Salem, a village in Dubai with a population of 589 residents, accounting for 0.02% of the total population of Dubai, as reported by the Dubai Statistic

Centre in 2018 [37]. The meteorological station is surrounded by the Marmoom Desert Conservation Reserve, which constitutes 10% of the total area of Dubai. It is the first unfenced desert reserve in the UAE. The total area exceeds 40 hectares of shrubland and contains many lakes with a total area of around 10 km^2 [38]. The amount of vegetation in the areas surrounding the three stations is negligible.

In Table 1, the population density of the three locations is calculated as the ratio between the population and the land area of the sectors of Dubai to which the weather stations belong. Considering as references the population densities of Sector 1 (i.e., 10,561/km^2) classified as "High" and of the Sector 9 (i.e., 5/km^2) classified as "Low", then Sector 2, the sector where the weather station of Dubai International Airport is located, is classified as an urban area with a "Medium" population density (3611/km^2). Al Maktoum International Airport is instead located within Sector 5, presenting a population density of 825/km^2 and therefore classified as a suburban area with a "Medium/Low" population density. The area of Saih Al-Salem presents a low population density of 7/km^2 and being located within Sector 9, was defined as a rural area for this study [39].

Table 1. Geographical information and description of the urban, suburban, and rural meteorological stations.

Weather Station No.	Weather Station Name	Latitude (°N)	Longitude (°E)	Station Elevation (m)	Surrounding Area	Population Density (pop./km^2)
1	Dubai International Airport	25°15′10″	55°21′52″	19	Urban	Medium (3611)
2	Al Maktoum International Airport	24°55′06″	55°10′32″	52	Suburban	Medium/Low (825)
3	Saih Al Salem	24°49′39″	55°18′43″	80	Rural	Low (7)

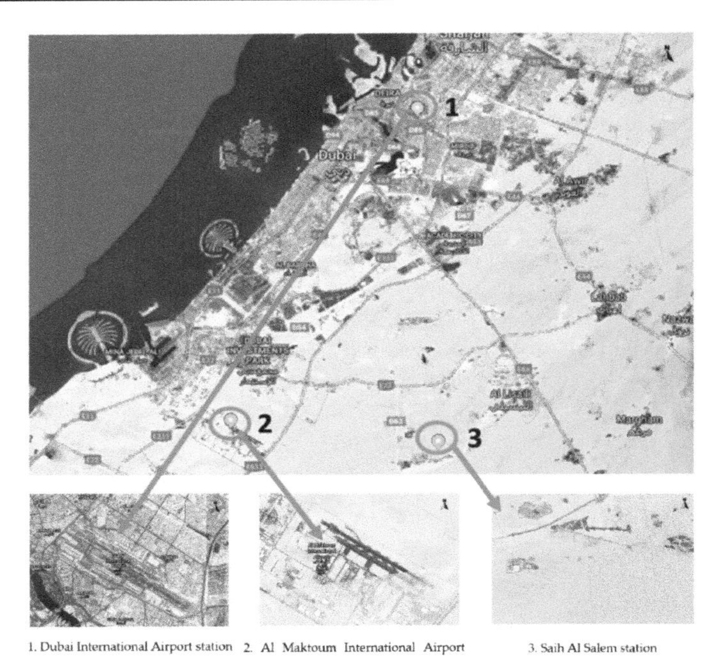

Figure 1. Location of the weather stations in Dubai as shown in Google Maps (1. Dubai International Airport station, 2. Al Maktoum International Airport station, 3. Saih Al Salem station).

2.2.1. Cluster Analysis of Climate Data

Clustering techniques were used to investigate the level of correlation between the microclimate parameters, i.e., air temperature and wind speed, monitored in the urban and suburban locations, and the canopy UHI intensity under the different ranges of wind directions. The cluster analysis was performed in the MS Excel tool only for the urban and suburban locations because those are the areas where the urbanization may produce an impact on the UHI intensity, while the rural area was considered as the reference location for the calculation of the UHI intensity. The data collected by each meteorological station for the entire period of interest (i.e., 5 years) were divided into six clusters based on the directions of the wind (i.e., either from the seaside or from the desert side, or from the coastal side which is in between, and divided according to four wind direction ranges). The characteristics of the six clusters are summarized in Table 2 and Figure 2.

Table 2. Clusters of wind direction.

Cluster No.	Wind Direction (°)	Directions Specification	No. of Measurements Dubai Airport	No. of Measurements Al Maktoum Airport
1	260–330	The sea	13,689	13,305
2	70–140	The desert	7873	7805
3	200–260	Coastal area	4407	4067
4	140–200	The desert	8848	9320
5	330–20	The sea	4298	4862
6	20–70	Coastal area	4709	4465

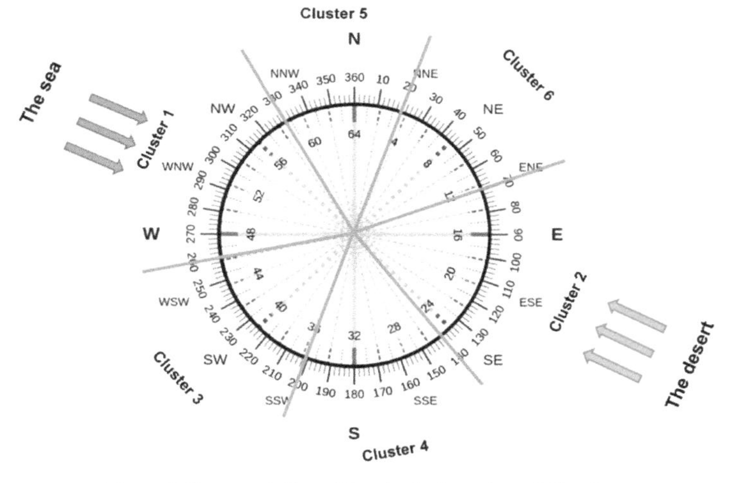

Figure 2. Representation of the six clusters of wind direction.

In Table 2, it can be observed that the dimensions of the six clusters (i.e., number of measurements included in each cluster) are consistent between the urban (i.e., station 1) and the suburban (i.e., station 2) area due to their close proximity. From the cluster analysis, the resulting predominant wind is the one coming from the sea, mainly seen in cluster 1 (i.e., 13,689 and 13,305 for station 1 and station 2, respectively) and cluster 5 (i.e., 4298 and 4862 for station 1 and station 2, respectively). The second predominant wind is the one coming from the desert, mainly seen in cluster 2, but also in cluster 4.

2.2.2. Canopy UHI Magnitudes

The hourly canopy UHI magnitude was calculated at 10 m above the ground for the urban and suburban locations for a period of five years (i.e., 2014–2018). Saih Al Salem (i.e., station 3) was chosen as a reference station, because it is located in a rural area not too far from the downtown city and close to the desert. Of the three stations, this was the least affected by the UHI due to its low level of urbanization. Since this reference station is located outside of the built-up area, it has natural desert coverage and negligible anthropogenic heat.

The canopy UHI magnitude (UHI intensity $_{T1-T3}$) was determined as the air temperature difference between Dubai International Airport (T1 Urban Average) which is station 1 and Saih Al Salem (T3 Rural Average) which is station 3, according to the following equation [40], Equation (1):

$$\text{UHI intensity}_{T1-T3}\ [^\circ\text{C}] = \text{T1 Urban Average}\ [^\circ\text{C}] - \text{T3 Rural Average}\ [^\circ\text{C}] \quad (1)$$

While the canopy UHI magnitude (UHI intensity $_{T2-T3}$) was determined by calculating the air temperature difference between Al Maktoum International Airport (T2 Suburban Average) which is station 2 and Saih Al Salem (T3 Rural Average) which is station 3, according to the following equation, [40] Equation (2):

$$\text{UHI intensity}_{T2-T3}\ [^\circ\text{C}] = \text{T2 Suburban Average}\ [^\circ\text{C}] - \text{T3 Rural Average}\ [^\circ\text{C}] \quad (2)$$

The canopy UHI calculations, performed according to the above equations, and the data post-processing presented in the result section, were based on the climate data analyzed in MS Excel.

3. Results

3.1. Microclimate Analysis

This section presents the microclimate analysis for the three investigated locations in Dubai.

Table 3 summarizes the results of the statistical analysis performed on the main weather parameters (i.e., air temperature, relative humidity, and wind speed) recorded by the three urban, suburban, and rural meteorological stations for the entire period of investigation (i.e., 2014–2018).

Table 3. Statistical data of the weather parameters for the three meteorological stations for 5 years (i.e., 2014–2018).

Weather Station No.	Weather Station Name	Weather Parameters	Max Value	Min Value	Average Value	Standard Deviation
1	Dubai International Airport	Temperature dry (°C)	48.6	12.3	29.6	6.6
		Relative humidity (%)	100	4	50	18.0
		Wind speed (km/h)	63	0	13	6.5
2	Al Maktoum International Airport	Temperature dry (°C)	48.5	7.1	28.0	7.8
		Relative humidity (%)	100	2	53	22.7
		Wind speed (km/h)	67	0	14	7.9
3	Saih Al Salem	Temperature dry (°C)	50.8	4.7	28.3	8.9
		Relative humidity (%)	100	1	48	26.0
		Wind speed (km/h)	67	0	10	7.0

Considering the entire 5 year period from 2014 to 2018, the average, maximum, and minimum air temperatures (measured at 10 m above the ground) were 29.6 °C, 48.6 °C, and 12.3 °C for Dubai International Airport station 1, 28.0 °C, 48.5 °C, and 7.1 °C for Al Maktoum International Airport station 2, and 28.3 °C, 50.8 °C, and 4.7 °C for Saih Al Salem station 3, respectively. Station 3 (i.e., Saih Al Salem station), located nearest to the desert, showed the lowest minimum and the highest maximum temperatures (i.e., highest thermal excursion) compared to the other two stations which, being located near the sea, present a local microclimate that is influenced by the Arabic Gulf. The minimum temperatures were recorded by station 3 in February 2014 and 2018 and were lower by about 7.6 °C and 2.4 °C than those recorded by Dubai Airport station and Al Maktoum Airport station, respectively. In July 2014 and 2017, the maximum temperatures recorded by station 3 were higher by about 2.2 °C and 2.3 °C than those recorded by the stations at Dubai Airport and Al Maktoum Airport, respectively.

Figure 3 summarizes, with the box plot representation, the air temperature measured by the three weather stations for the entire period of investigation (i.e., 2014–2018) and for each year separately.

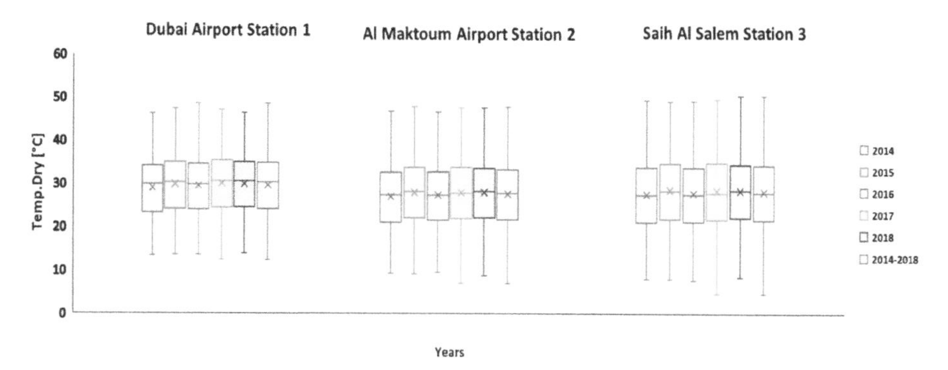

Figure 3. Box plot of the air temperature for the urban, suburban, and rural areas in Dubai for 5 years (i.e., 2014–2018).

As expected, each station has an air temperature profile slightly different from the other two, despite their proximity to each other. The different locations of the stations with respect to the desert, the built area, and the coast, produced significant differences in terms of air temperature. The synoptic conditions (i.e., the hot air coming from the desert, the cool air coming from the sea or sea breeze) that characterize the entire area interact in a different way in the proximity of each meteorological station.

According to the collected relative humidity data, which affects the local microclimate, the Summer season is characterized by around 90% humidity. The maximum value recorded in all three stations is 100%, while the minimum and average values range from 1% to 4% and from 48% to 53%, respectively, where as expected, the highest values are experienced in the urban and suburban areas that are located closer to the sea, with a negative impact on human activities and outdoor thermal comfort.

A limited presence of clouds was observed in the urban and suburban areas of Dubai during the monitored period, with a clear annual and seasonal variability of the cloud coverage. In the 5 years of monitoring, the cloud cover rarely reached the maximum value of 9 octas (i.e., sky obstructed from view). Only for 14% and 12% of the time, the sky was between 4 octas (i.e., sky half cloudy) and 9 octas during the monitoring period in the urban and suburban areas, respectively.

In contrast, the total absence of cloud cover (i.e., 0 octas) was recorded very frequently along the monitored period (i.e., 67% and 69% of the time in the urban and suburban areas, respectively) leading to high values of the incident solar radiation, and as a result, to a large amount of the heat absorbed by the built environment during the daytime. This absorbed heat is then released into the atmosphere during the night time. Under these conditions, the nocturnal radiative cooling will be less effective in the built areas with respect to the desert rural area that easily cools down. This contributes to the night-time UHI phenomenon, where the temperature difference between the urban and rural areas is positive at night.

When the sky is completely obstructed from view (i.e., 9 octas), the night-time UHI intensity $_{T1–T3}$ and $_{T2–T3}$ reach the maximum (minimum) value of 7.1 °C (0.3 °C) and 3.4 °C (−0.9 °C), respectively, when measured in correspondence to station 1, and of 9.4 °C (0.3 °C) and 7.8 °C (−3.0 °C), respectively, when measured in correspondence to station 2. When the sky is completely clear (i.e., 0 octas), the night-time UHI intensity $_{T1–T3}$ and $_{T2–T3}$ reach the maximum (minimum) value of 11.5 °C (−3.3 °C) and 13.0 °C (−7.7 °C), respectively, when measured in correspondence to station 1, and of 11.5 °C (−3.3 °C) and 13.0 °C (−8.9 °C), respectively, when measured in correspondence to station 2. Thus, the night time UHI variability range is higher under clear sky conditions (i.e., 0 octas) than under cloud-covered sky conditions (i.e., 9 octas). The same result is obtained for the UHI intensity value, as can be observed in Figure 4, which shows the existing relation between the UHI intensity $_{T1–T3}$ and $_{T2–T3}$ and the cloud cover for the entire monitored period of 5 years and for the urban and suburban areas.

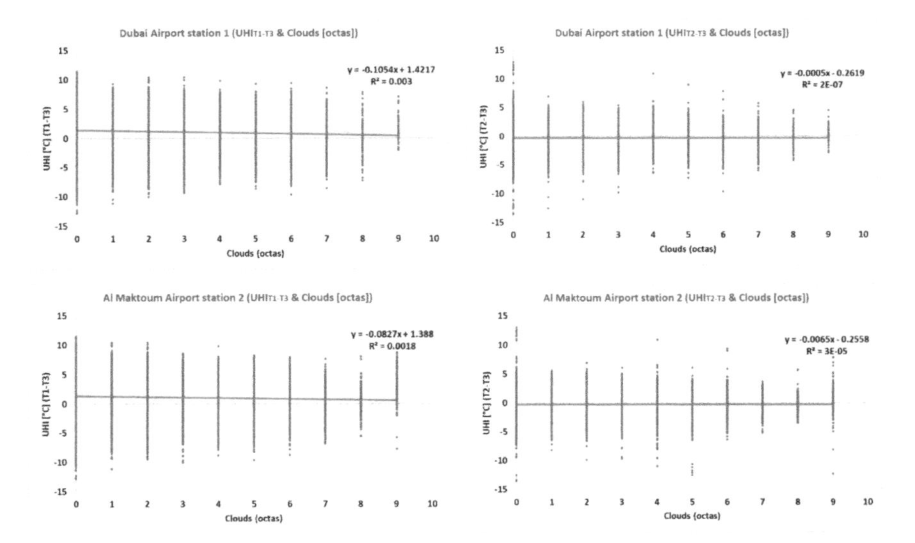

Figure 4. Comparison between the urban heat island (UHI) and the cloud cover relation: UHI intensity $_{T1–T3}$ and cloud cover from station 1 (**top left**), UHI intensity $_{T2–T3}$ and cloud cover from station 1 (**top right**), UHI intensity $_{T1–T3}$ and cloud cover from station 2 (**bottom left**), UHI intensity $_{T2–T3}$ and cloud cover from station 2 (**bottom right**).

3.2. *Daytime and Night-Time UHI Intensity and Frequency Distribution*

Figure 5 shows the daytime (i.e., 6 a.m.–9 p.m.) and night-time (i.e., 9 p.m.–6 a.m.) boxplot of the canopy UHI intensity $_{T1–T3}$ and $_{T2–T3}$ for the entire period of five years considered in this study.

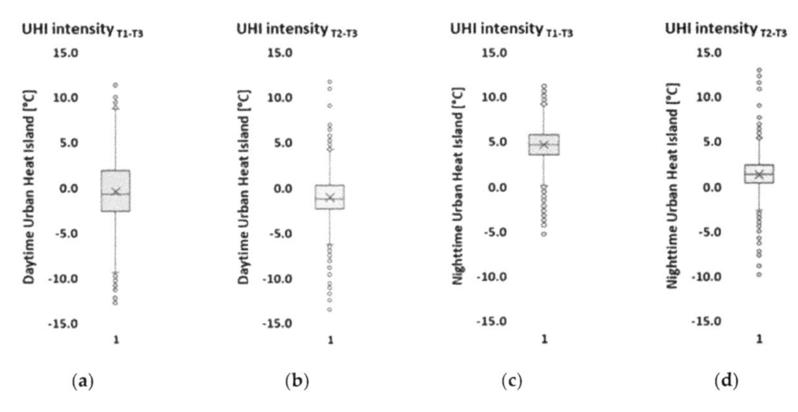

Figure 5. Box plot of the daytime and night-time UHI intensity for 5 years (i.e., 2014–2018): (**a**) daytime UHI intensity $_{T1–T3}$, (**b**) daytime UHI intensity $_{T2–T3}$, (**c**) night-time UHI intensity $_{T1–T3}$, (**d**) night-time UHI intensity $_{T2–T3}$.

The daytime and night-time UHI $_{T2-T3}$ show higher maximum values of 11.7 °C and 13 °C compared to the daytime and night-time UHI $_{T1-T3}$ values which are 11.3 °C and 11.5 °C, respectively (Figure 5 and Table 4). The average and median values recorded for the daytime and night-time UHI intensities $_{T1-T3}$ and $_{T2-T3}$ are shown in Table 4 together with the maximum and minimum values.

Table 4. Daytime and night-time UHI intensities $_{T1-T3}$ and $_{T2-T3}$ values for 5 years (i.e., 2014–2018).

Value Discretion	UHI Intensity T1–T3 (°C)	Daytime UHI Intensity T1–T3 (°C) (6 a.m.–9 p.m.)	Night-time UHI Intensity T1–T3 (°C) (9 p.m.–6 a.m.)	UHI Intensity T2–T3 (°C)	Daytime UHI Intensity T2–T3 (°C) (6 a.m.–9 p.m.)	Night-time UHI Intensity T2–T3 (°C) (9 p.m.–6 a.m.)
Maximum value	11.5	11.3	11.5	13.0	11.7	13.0
Minimum value	−12.8	−12.8	−5.4	−13.5	−13.5	−9.9
Average value	1.3	−0.4	4.6	−0.3	−1.1	1.3
Median value	1.7	−0.7	4.6	−0.2	−1.3	1.4

It is worth noting that, considering the period of interest and as summarized in Table 4, the urban and suburban areas of Dubai were in average warmer at night than the rural area by 4.6 °C and 1.3 °C, respectively. On the contrary, during the day, the rural area was in average warmer than the urban and suburban areas (i.e., 0.4 °C and 1.1 °C, respectively).

The frequency distribution of the daytime and night-time values for the UHI intensity $_{T1-T3}$ and $_{T2-T3}$ given in (Figure 6) help to understand the distribution of the values representing the UHI intensity during the period of interest.

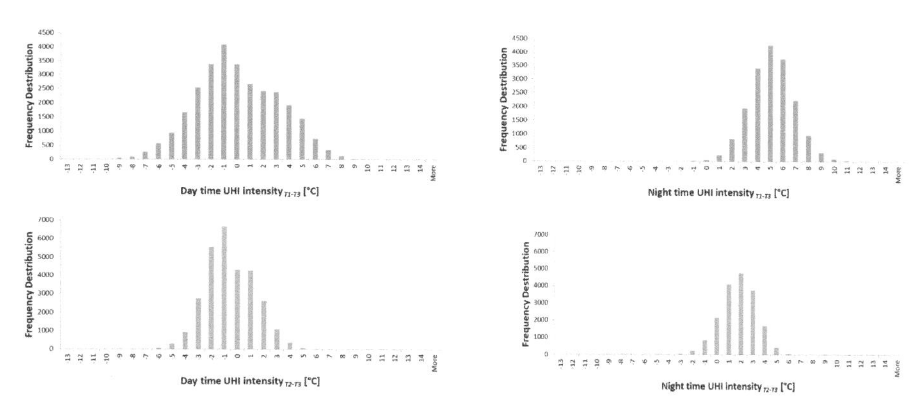

Figure 6. Frequency distribution of the daytime and night-time UHI intensity for 5 years (i.e., 2014–2018): daytime UHI intensity $_{T1-T3}$ (**top left**), night-time UHI intensity $_{T1-T3}$ (**top right**), daytime UHI intensity $_{T2-T3}$ (**bottom left**), night-time UHI intensity $_{T2-T3}$ (**bottom right**).

Figures 5 and 6 show that the UHI intensity is higher at night than during the daytime in both urban and suburban areas. The urban area presents higher values of average UHI intensity with respect to the suburban area both during the night (i.e., 3.3 °C of difference) and the day (i.e., 0.7 °C of difference).

3.3. *UHI Intensity* $_{T1-T3}$ *and Climate Parameters Collected by Station 1*

The air temperature and wind speed measured by station 1, located in the Dubai International Airport, were investigated for the six clusters of different ranges of wind direction in relation with the canopy urban heat island intensity (UHI) calculated as the difference between the air temperature measured at the Dubai International Airport (i.e., station 1) and at Saih Al Salem (i.e., station 3), where station 1 is considered as located in an urban area and station 3 is considered as located in a suburban area (Figure 7).

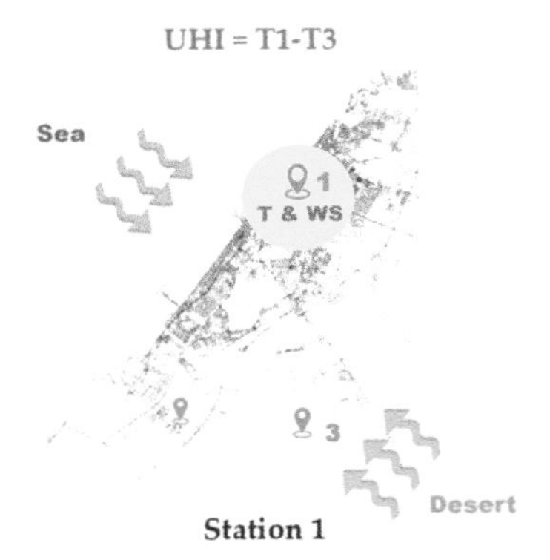

Figure 7. Representation of the UHI calculation and the station (i.e., urban station under the yellow circle) selected for the investigation of the climate parameters, i.e., air temperature (T) and wind speed (WS).

Figure 8 shows a comparison of the relation between the hourly canopy UHI intensity (UHI intensity $_{T1-T3}$) and the hourly air temperature measured by station 1 under a different cluster of wind directions. A moderate negative correlation between the temperature and the UHI intensity was found for cluster 1 (i.e., wind direction between 260° and 330° from the seaside) (Figure 8a) where the UHI ranges between 11.5 °C and −12.8 °C, cluster 3 (i.e., wind direction between 250° and 210° from the coastal area) (Figure 8c) where the UHI ranges between 10.1 °C and −7.8 °C, and cluster 4 (i.e., wind direction between 200° and 150° from the desert) (Figure 8d) where the UHI ranges between 11.4 °C and −11.2 °C. The temperature for these three clusters fluctuates between 47 °C and 12.3 °C. A weak negative correlation between the temperature and UHI intensity was found for all the remaining clusters except for cluster 5 (i.e., wind direction located between 20° and 340°) (Figure 8e) where no correlation was found. The results show that the UHI intensity varies with different wind directions. When the wind is blowing from the desert (i.e., cluster 2) (Figure 8b), the temperature and the UHI are almost independent. Despite the positive impact of the sea breeze, coastal cities suffer from the UHI [41]. Different experimental and numerical investigations have shown the impact of the sea on the development of the UHI in coastal cities [42,43]. The temperature in station 1 is affected by UHI when the wind is blowing from the seaside. It seems that, when there is a sea breeze combined with high temperatures, the negative UHI value is high (Figure 8a). This is due to the fact that the high temperatures recorded by station 1 are mitigated by the sea breeze more than the temperatures recorded by rural station 3, which is closer to the desert and typically reaches higher temperatures during hot days. This pattern is due to advection from the sea breeze cooling mechanism and the flow of air, which is affected by buildings. These results align with those of similar studies in Athens and Sydney [2,7,44,45].

Figure 9 shows a comparison of the relation between canopy UHI intensity (UHI intensity $_{T1-T3}$) and the wind speed measured by station 1 under different clusters of wind directions.

For the wind speed, as for the air temperature, a negative slope of the regression line (p-value < 0.05) is evident in all clusters except in cluster 5 (Figure 9e) where there is no correlation. In this case, the correlation coefficient (R) between the UHI and the wind speed is negative and weak (ranging between −0.2–−0.4 for all clusters) regardless of the wind direction. The results show that the wind speed is lower when the wind blows from the coastal area (i.e., clusters 3 and 6) (Figure 9c,f) with average wind speeds of 10 km/h and 11 km/h, respectively. It reaches higher values when the wind blows from the sea (i.e., cluster 1) (Figure 9a) with average wind speeds of about 18 km/h due to the energy produced by the sea breeze. The results show that wind speed has an impact on the magnitude of the UHI. Results show that the low values of wind speed are more conducive to the development of a UHI [46]. Thus, the data from station 1 shows an inverse relation between the wind speed and UHI intensity when the wind speed increases for all clusters except cluster 5. This relation is consistent with several studies conducted in various regions [2,47–49]. In cluster 5 (Figure 9e), with the wind from the North, no correlation was found between the wind speed and UHI intensity [2].

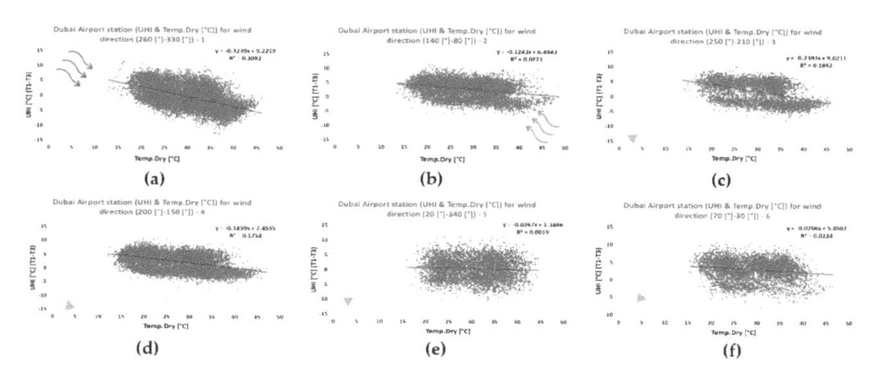

Figure 8. Comparison between the UHI and the temperature relation under different clusters of wind directions: (**a**) cluster 1, (**b**) cluster 2, (**c**) cluster 3, (**d**) cluster 4, (**e**) cluster 5, and (**f**) cluster 6.

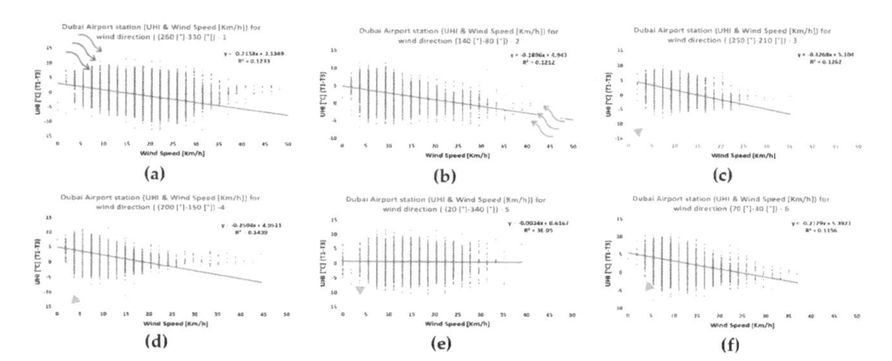

Figure 9. Comparison between the UHI and the wind speed relation under different clusters of wind directions: (**a**) cluster 1, (**b**) cluster 2, (**c**) cluster 3, (**d**) cluster 4, (**e**) cluster 5, and (**f**) cluster 6.

3.4. *UHI Intensity* $_{T1-T3}$ *and Climate Parameters Collected by Station 2*

The air temperature and wind speed measured by station 2 were investigated for the six clusters of wind direction in relation to the UHI calculated as in the previous section and as shown in Figure 10.

A strong negative correlation between the air temperature measured by station 2 and (UHI intensity $_{T1-T3}$) was found for cluster 1 (Figure 11a) and cluster 3 (Figure 11c), where the UHI values range from 10.7 °C to −12.8 °C and the temperatures range between 48.2 °C and 12.4 °C for both clusters. The values of the correlation coefficient comprise about 40% of the data. Regarding the other clusters, there is a weak and negative linear relation between the temperature and the UHI intensity as shown in Figure 11.

As evident in Figure 12, the canopy UHI intensity has a moderate negative correlation with wind speed, averaging around 14 km/h, in cluster 5 (Figure 12e) where the wind comes from the seaside direction. The other clusters in the same station show a weak and negative correlation with close (R) values varying between −0.2 and −0.5. The results indicate that the high values of wind speed decrease the differences in the temperatures for all clusters, as observed in the study of Sydney [2] as well as in [50]. Additionally, the impact on UHI intensity in terms of all wind directions is minimized due to the effects of the two parameters. An increase in wind speed in the urban area enhances the heat flux, whose heating impact on the urban environment is stronger than the enhanced cooling effect [49,51]. It could also be that an urban atmosphere can absorb the heat from solar radiation during the daytime. These outcomes are supported by the results of previous studies [52,53]. As shown, the station recorded a maximum ambient temperature of 48.5 °C in July 2015, while the reference station recorded 50.2 °C for the same year.

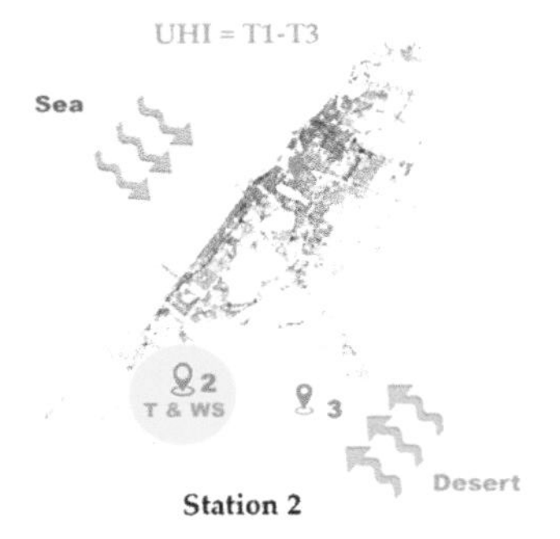

Figure 10. Representation of the UHI calculation and the station (i.e., the suburban station under the yellow circle) selected for the investigation of the climate parameters, i.e., air temperature (T) and wind speed (WS).

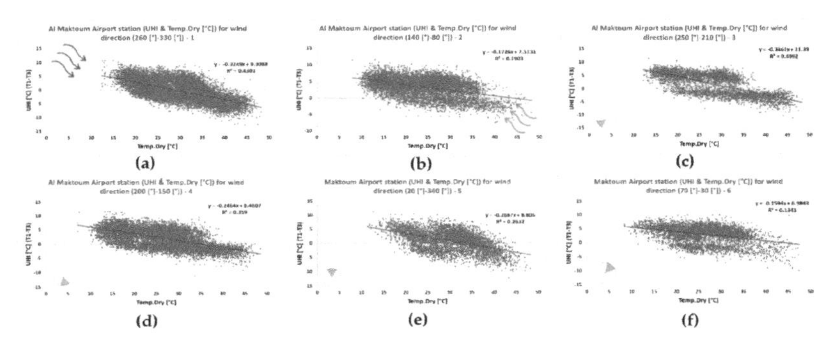

Figure 11. Comparison between the UHI and the temperature relation under different clusters of wind directions: (**a**) cluster 1, (**b**) cluster 2, (**c**) cluster 3, (**d**) cluster 4, (**e**) cluster 5, and (**f**) cluster 6.

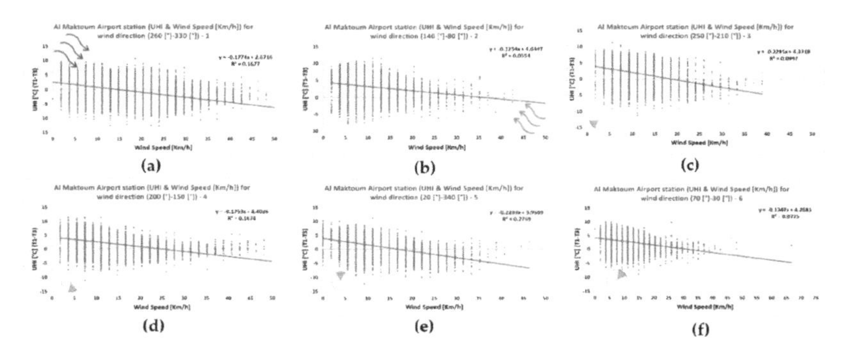

Figure 12. Comparison between the UHI and the wind speed relation under different clusters of wind directions: (**a**) cluster 1, (**b**) cluster 2, (**c**) cluster 3, (**d**) cluster 4, (**e**) cluster 5, and (**f**) cluster 6.

3.5. UHI Intensity $_{T2-T3}$ and Climate Parameters Collected by Station 1

The air temperature and wind speed measured by station 1 were investigated for the six clusters of different ranges of wind direction in relation with the canopy urban heat island intensity (UHI) calculated as the difference between the air temperature measured at the Al Maktoum International Airport (i.e., station 2) and at Saih Al Salem (i.e., station 3) (Figure 13).

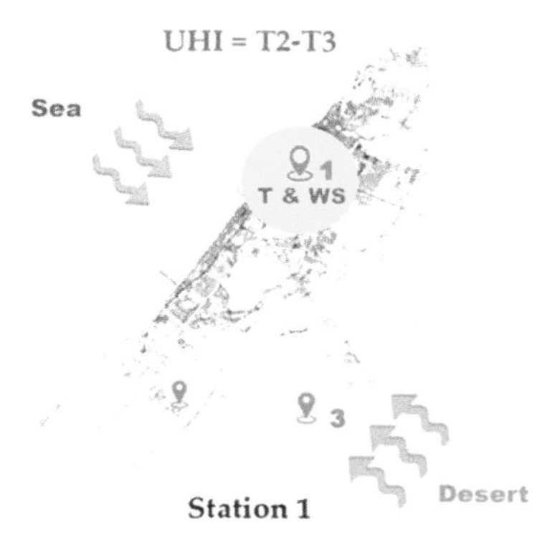

Figure 13. Representation of the UHI calculation and the station (i.e., the urban station under the yellow circle) selected for the investigation of the climate parameters, i.e., air temperature (T) and wind speed (WS).

Figure 14 shows a moderate negative linear relation between the temperature and the (UHI intensity $_{T2-T3}$) for cluster 1 (Figure 14a). The temperature for this cluster ranges between 46.1 °C and 13 °C with an average value close to 30.8 °C and the UHI ranges from 12.4 °C to −13.5 °C. For the five other clusters, the canopy UHI intensity shows a weak negative correlation with temperature, with few variations. Here, the synoptic climate conditions associated with the advection and convection phenomena play a role in this mechanism as does the additional anthropogenic heat in the area. These findings are supported by several other studies [2,7,54,55]. Other previous studies have reported that the daytime intensity of the heat island is reduced under specific synoptic conditions [56,57].

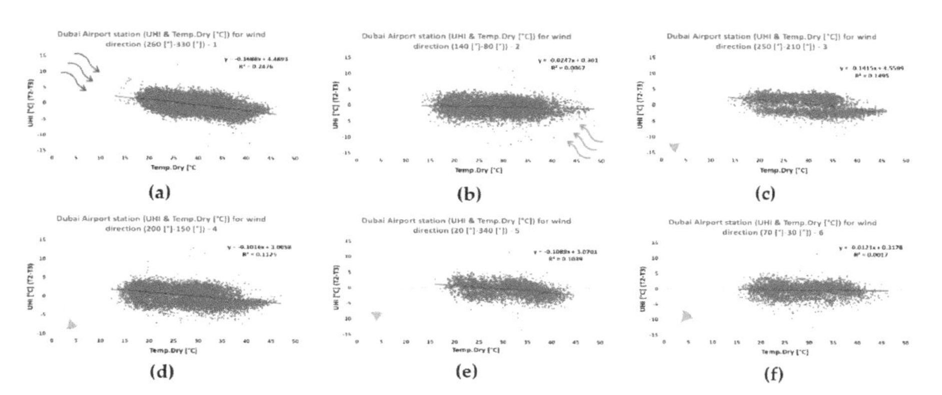

Figure 14. Comparison between the UHI and the temperature relation under different clusters of wind directions: (**a**) cluster 1, (**b**) cluster 2, (**c**) cluster 3, (**d**) cluster 4, (**e**) cluster 5, and (**f**) cluster 6.

In Figure 15, the (UHI intensity $_{T2-T3}$) ranges between 11.6 °C and −13.4 °C showing a moderate negative correlation when the wind speed ranges from 39 km/h to 0 km/h, with an average value close to 15 km/h when the wind is coming from the North as in cluster 5 (Figure 15e). The data of the cluster demonstrate a 22.8% variation in the UHI intensity compared with the other clusters which have a very weak correlation with the UHI intensity as they are close to zero. Moreover, in regard to western and north-western wind directions, there is a negative relation between the UHI and both the temperature and wind speed, similar to the findings of other studies [2].

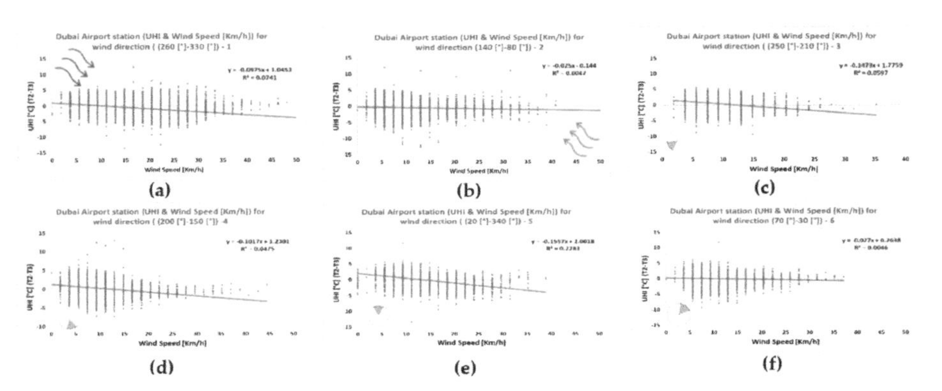

Figure 15. Comparison between the UHI and the wind speed relation under different clusters of wind directions: (**a**) cluster 1, (**b**) cluster 2, (**c**) cluster 3, (**d**) cluster 4, (**e**) cluster 5, and (**f**) cluster 6.

3.6. UHI Intensity $_{T2–T3}$ and Climate Parameters Collected by Station 2

The air temperature and wind speed measured by station 2 were investigated for the six clusters of wind direction in relation to the UHI calculated as described in the previous section and as shown in Figure 16.

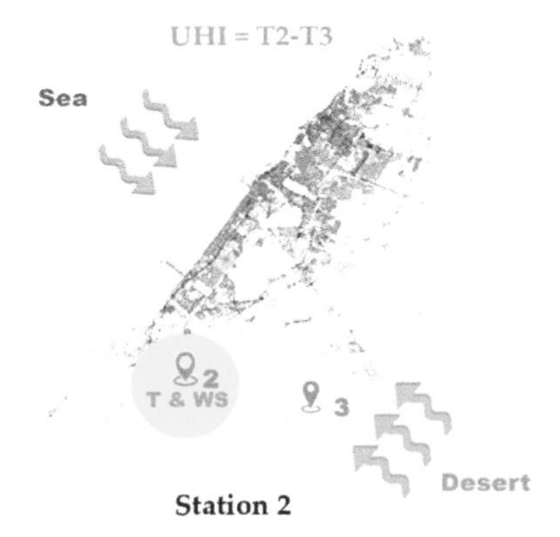

Figure 16. Representation of the UHI calculation and the station (i.e., the suburban station under the yellow circle) selected for the investigation of the climate parameters, i.e., air temperature (T) and wind speed (WS).

Cluster 1 (Figure 17a) and cluster 3 (Figure 17c) show a moderate negative correlation between the temperature and (UHI intensity $_{T2–T3}$). The temperature in both clusters ranges between 48.2 °C and 12.2 °C. However, in the other clusters, there is a negative weak correlation between the temperature and the UHI intensity, as shown in Figure 17.

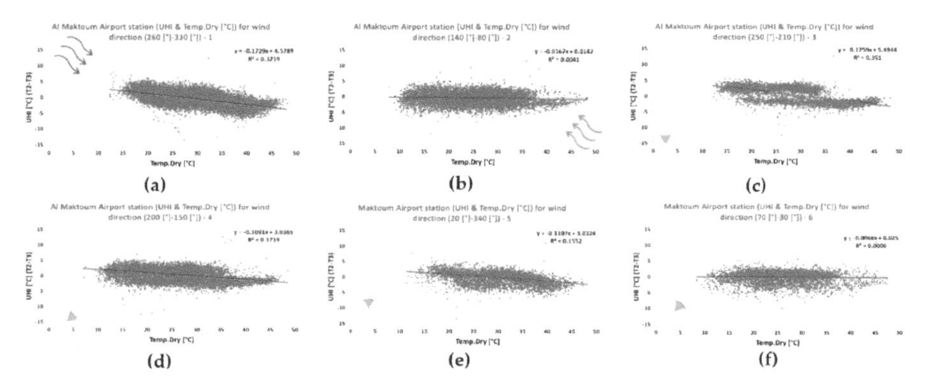

Figure 17. Comparison between the UHI and the temperature relation under different clusters of wind directions: (**a**) cluster 1, (**b**) cluster 2, (c) cluster 3, (**d**) cluster 4, (**e**) cluster 5, and (**f**) cluster 6.

Figure 18 shows a weak negative correlation which means little relation between the wind speed and UHI intensity; it is close to zero in all clusters [2] for Al Maktoum International Airport (station 2). The results show no significant difference between station 1 and station 2 regarding the climate parameters and the UHI intensity phenomenon. In the surrounding area of station 1, the prevailing wind is coming from the sea. It is clear that when the wind is coming from the seaside or the desert, due to the higher energy embedded in the synoptic conditions, the wind speed will be higher than when the wind direction is parallel to the coast (cluster 3) (Figure 18c) These conclusions are supported by previous studies undertaken in Asian and Australian cities [2,40,49,58].

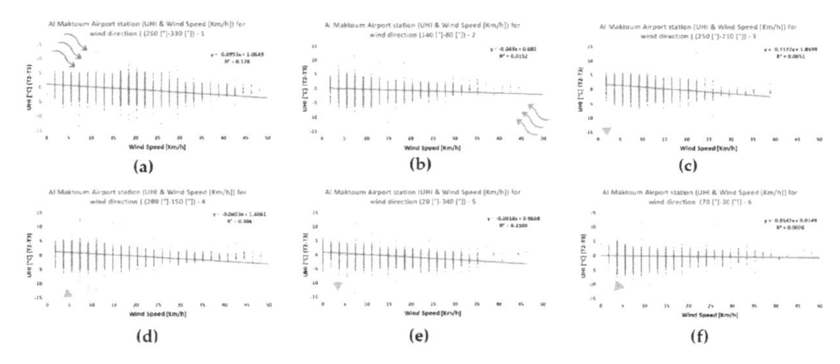

Figure 18. Comparison between the UHI and the wind speed relation under different cluster of wind directions: (**a**) cluster 1, (**b**) cluster 2, (c) cluster 3, (**d**) cluster 4, (**e**) cluster 5, and (**f**) cluster 6.

3.7. *UHI Intensity $_{T1-T3}$ and Climate Parameters Collected by Station 3*

As done previously with both stations 1 and 2, the same procedure was followed to determine the temperature and wind speed for Saih Al Salem (i.e., station 3) and their relation with the UHI magnitude, calculated as the difference between the temperature measured at the Dubai International Airport (i.e., station 1) and at the Saih Al Salem area for all clusters of wind direction (Figure 19).

Figure 20 shows that for all clusters, there is a strong negative correlation between the UHI intensity $_{T1-T3}$, ranging from 11.5 °C to −12.8 °C, and the air temperature, fluctuating between 50.8 °C and 4.7 °C, with an average of 28.3 °C. To recap, cluster 1 (Figure 20a) is from the sea direction, cluster 2 (Figure 20b) from the desert area, and cluster 3 (Figure 20c) from the coastal area. In terms of this parameter, there is a variation from 55% to 65% in UHI intensity among these three clusters. When the temperature increases, so does the intensity of the UHI.

Figure 19. Representation of the UHI calculation and the station (i.e., the rural station under the yellow circle) selected for the investigation of the climate parameters, i.e., air temperature (T) and wind speed (WS).

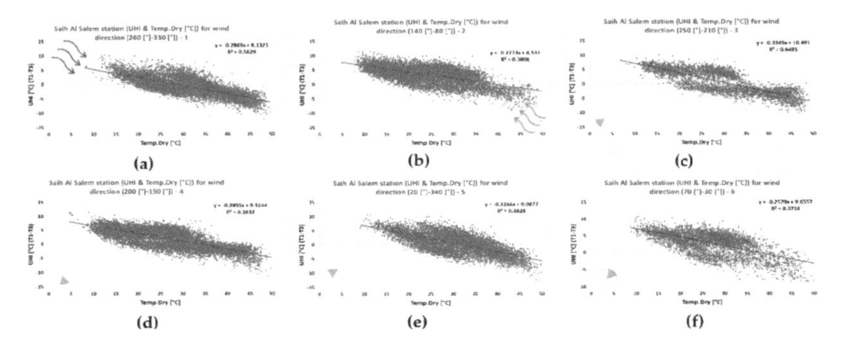

Figure 20. Comparison between the UHI and the temperature relation under different clusters of wind directions: (**a**) cluster 1, (**b**) cluster 2, (**c**) cluster 3, (**d**) cluster 4, (**e**) cluster 5, and (**f**) cluster 6.

The slope of the regression line (p value < 0.05) shown in Figure 21 indicates that for all clusters, there is a moderate negative relation between the wind speed and the UHI intensity. For all six clusters, the results indicate a relation ranging from moderate to strong between the UHI intensity and the two climate parameters in Saih Al Salem station. Synoptic weather conditions could have an impact on the parameter mechanisms as could the open desert space criterion. Numerous previous studies have reported the same results [2,47,54].

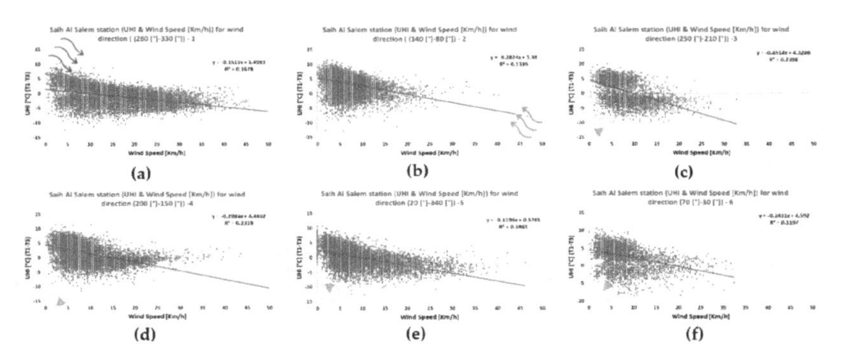

Figure 21. Comparison between the UHI and the wind speed relation under different clusters of wind directions: (**a**) cluster 1, (**b**) cluster 2, (**c**) cluster 3, (**d**) cluster 4, (**e**) cluster 5, and (**f**) cluster 6.

3.8. *UHI Intensity* $_{T2-T3}$ *and Climate Parameters Collected by Station 3*

The temperature and wind speed measured by station 3 were investigated for the six clusters of wind direction as in the previous sections. The UHI was calculated as the difference between the temperature measured by station 2 (i.e., Al Maktoum International Airport) and station 3 (i.e., Saih Al Salem) as shown in Figure 22.

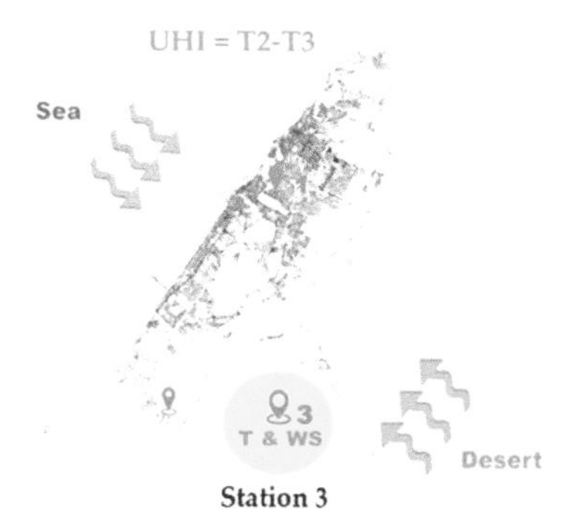

Figure 22. Representation of the UHI calculation and the station (i.e., the urban station under the yellow circle) selected for the investigation of the climate parameters, i.e., air temperature (T) and wind speed (WS).

The scatter plots in Figure 23 show strong negative correlations for the different clusters between the UHI intensity $_{T2-T3}$ and the temperature in cluster 1 (Figure 23a) from the sea direction, cluster 3 (Figure 23c) from the coastal area, cluster 4 (Figure 23d) from the desert side, and cluster 5 (Figure 23e) from the sea. The data demonstrates about 47%, 57%, 41%, and 37% variation in the canopy UHI, respectively. For the other two clusters, there is a weak and negative linear association between the temperature and UHI intensity; also, the temperature had independent effects on UHI intensity [2,7,55,59].

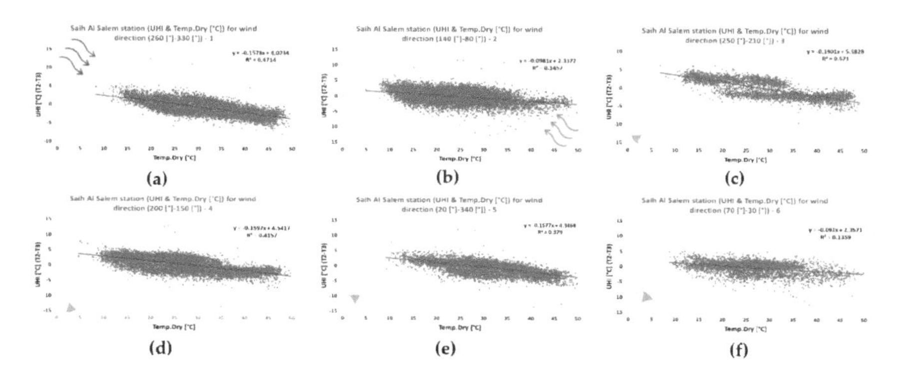

Figure 23. Comparison between the UHI and the temperature relation under different clusters of wind directions: (**a**) cluster 1, (**b**) cluster 2, (**c**) cluster 3, (**d**) cluster 4, (**e**) cluster 5, and (**f**) cluster 6.

As shown in Figure 24, the slope of the regression line (p value < 0.05) indicates a moderate negative correlation between the wind speed and the canopy UHI magnitude in cluster 3 (Figure 24c), cluster 4 (Figure 24d) and cluster 5 (Figure 24e). The UHI intensity for these three clusters is between 13 °C and −13.5 °C, while the wind speed ranges from 51 km/h to 0 km/h, with an average value close to 9 km/h. The results indicate moderate to strong relations between the UHI magnitude and the two climate parameters for this station. These results may be due to the presence of different synoptic weather systems of heating mechanism, a desert wind with a cooling mechanism, and coastal wind, which are created by the combination of climate parameters. These conclusions are compatible with those of previous studies conducted in various regions [2,47,55].

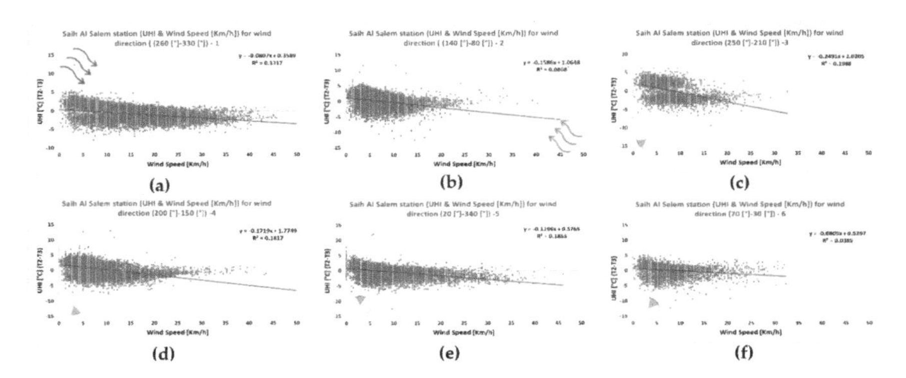

Figure 24. Comparison between the UHI and the wind speed relation under different cluster of wind directions: (**a**) cluster 1, (**b**) cluster 2, (**c**) cluster 3, (**d**) cluster 4, (**e**) cluster 5, and (**f**) cluster 6.

3.9. Correlation between UHI Intensity $_{T1-T3}$ and $_{T2-T3}$ and Climate Parameters

Figure 25 shows a comparison between the UHI intensity $_{T1-T3}$ and UHI intensity $_{T2-T3}$ and the hourly air temperature measured by station 1 and station 2 under the respective clusters of wind directions.

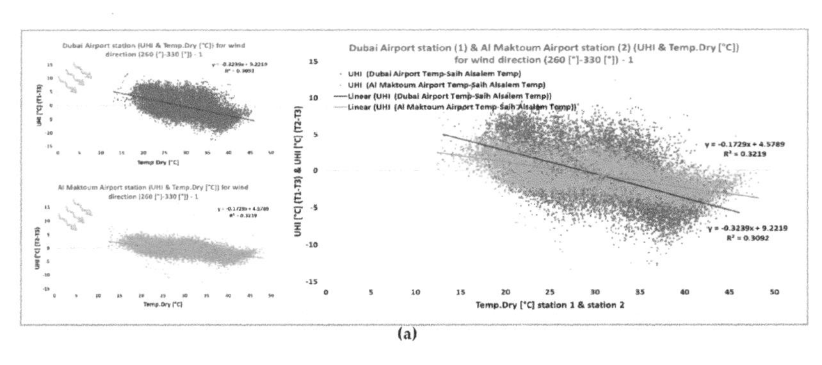

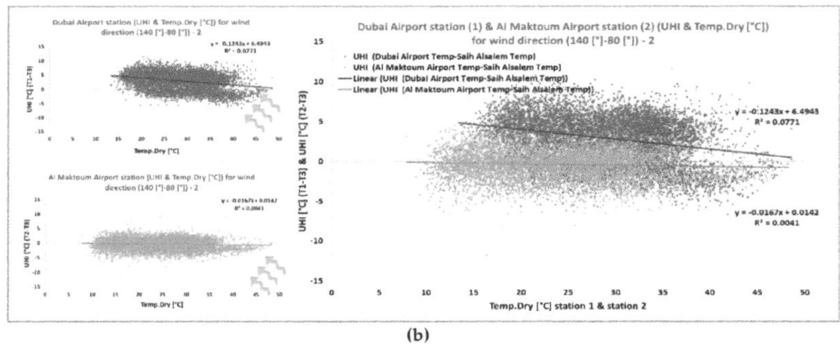

Figure 25. *Cont.*

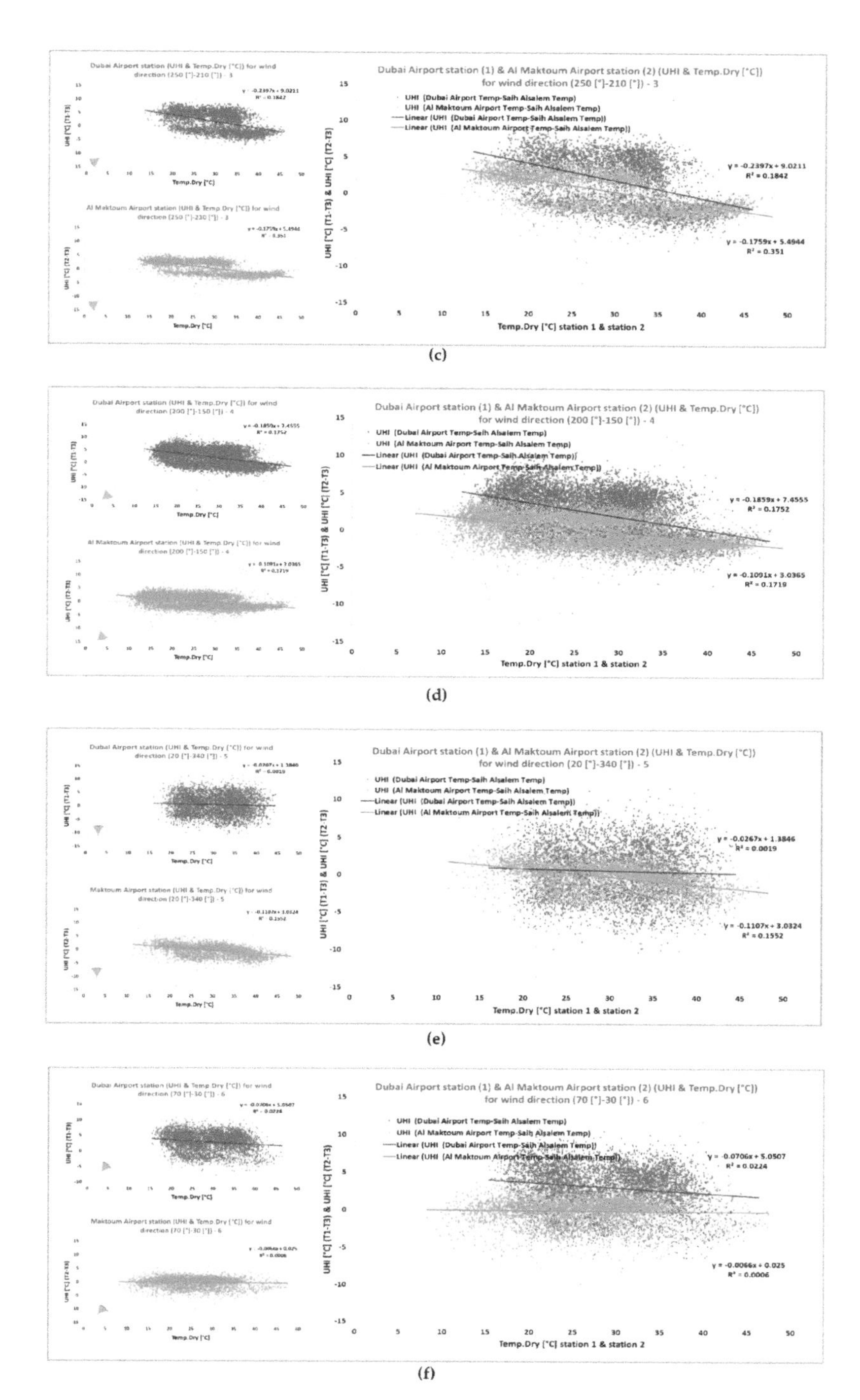

Figure 25. Comparison between the UHI and the air temperature relation measured in the urban (green) and suburban (orange) areas under different clusters of wind directions: (**a**) cluster 1, (**b**) cluster 2, (**c**) cluster 3, (**d**) cluster 4, (**e**) cluster 5, and (**f**) cluster 6.

A strong and negative correlation coefficient (R) between the temperature and UHI intensity $_{T1-T3}$ and $_{T2-T3}$ was found in Al Maktoum International Airport when the wind was coming from the sea. Hence, the temperature only demonstrates about 32% variation in the urban heat island intensity. Whereas, when the wind is coming from the desert side (cluster 2) (Figure 25b), there is no correlation between the temperature and the intensity of UHI at station 2 (i.e., Al Maktoum International Airport), which is close to zero. These results are aligned with those of other studies conducted in different regions [2,49,58].

Figure 26 compares the correlations between the wind speed and the canopy UHI intensity for both stations through the same procedure used for measuring the parameters of temperature and UHI intensity.

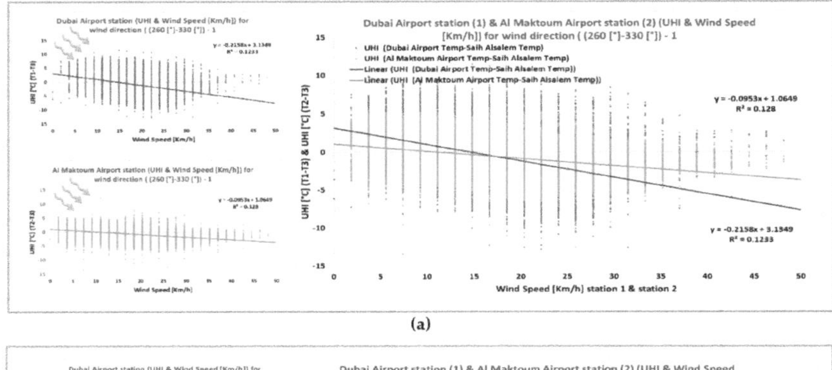

(a)

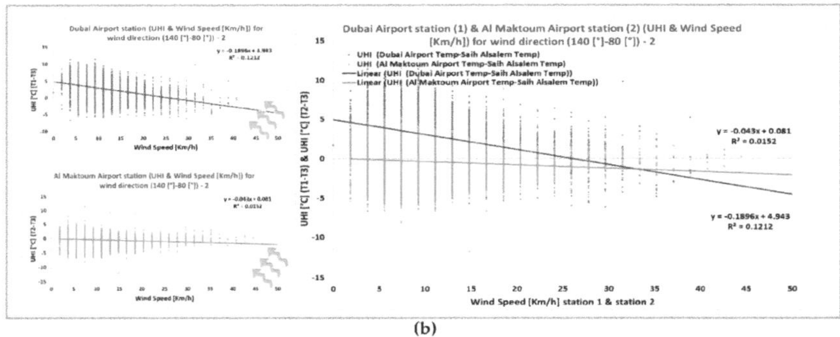

(b)

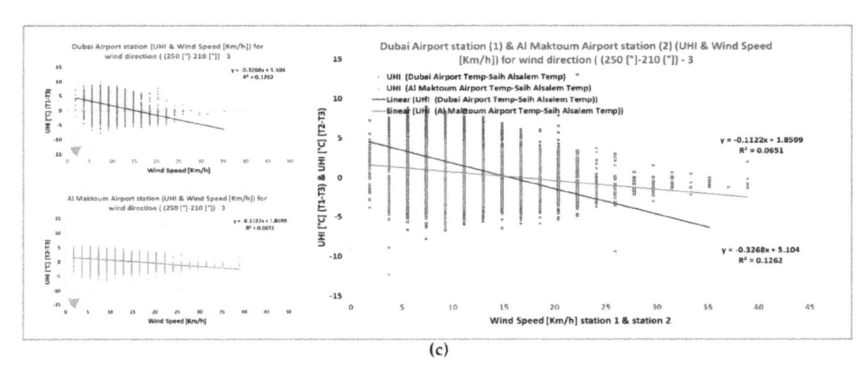

(c)

Figure 26. *Cont.*

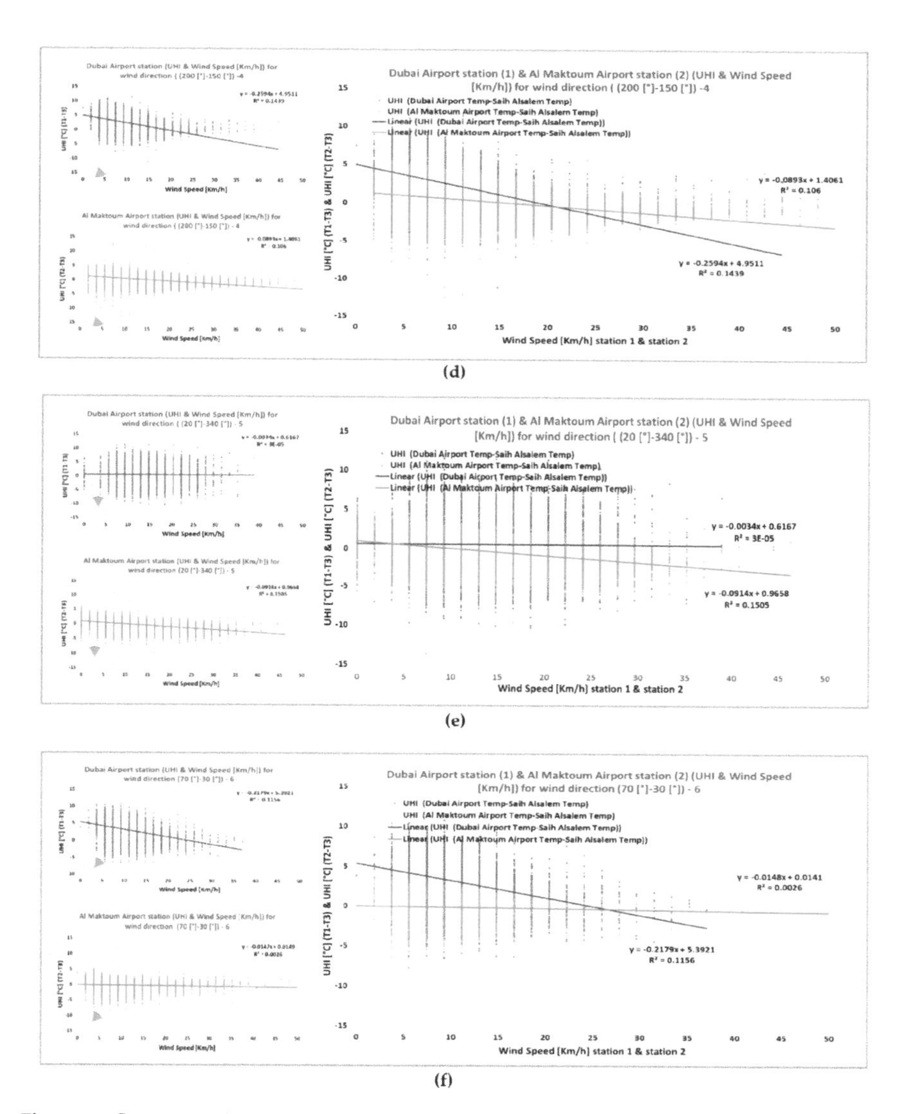

Figure 26. Comparison between the UHI and the wind speed relation measured in the urban (green) and suburban (orange) areas under different clusters of wind directions: (**a**) cluster 1, (**b**) cluster 2, (**c**) cluster 3, (**d**) cluster 4, (**e**) cluster 5, and (**f**) cluster 6.

Results show that when the wind direction is coming from the sea, there is a weak negative correlation between the wind speed and the UHI intensity. For cluster 1 (Figure 26a), the wind speed is responsible for only a 12% variation in the UHI magnitude. However, when the wind is coming from the desert side as in cluster 2 (Figure 26b), there is no correlation between the wind speed UHI intensity. These results are supported by those of other studies [2].

4. Discussion

Urban areas are facing several climate issues, one of which is the urban heat island (UHI) phenomenon. Thus, to understand the characteristics of UHI and its effects on the built environment, this study investigated the correlation between the canopy UHI and two meteorological parameters, air temperature and wind speed, with given specific ranges of wind direction. This was done by collecting and analyzing the hourly data from three meteorological stations situated in different areas of Dubai (i.e., urban, suburban, and rural areas).

It has been observed that the UHI intensity is larger at night in both urban and suburban areas and that the urban area presents the highest values of average UHI intensity. The main reason for the negative UHI at night is the presence of the heat that, given the intense incident solar radiation, remains trapped in the urban and suburban areas according to the geometry and density of the built environment and the thermal-optical properties of the urban surfaces. During the day, the built areas in Dubai act as urban cool islands showing negative UHI intensity values.

The analysis results show that canopy UHI intensity varies with wind direction and meteorological conditions. This has been confirmed by other studies carried out in cities with different climatic conditions [60,61]. Moreover, in regard to UHI intensity $_{T1-T3}$, the data from the urban meteorological station indicated that air temperature and UHI were almost independent when the wind was coming from the desert (i.e., from the East and South-East), while the temperature is affected by UHI when the wind is flowing from the sea (i.e., West and North-West) thanks to the cooling mechanism of the sea breeze and the short distance of the seashore to the station 1 [41,43]. It could also be influenced by the building's orientation. These results are consistent with the findings of studies conducted in Sydney where the eastern suburbs benefited from the significant cooling mechanism of the sea breeze and the variations in UHI were a function of the prevailing climate conditions [2]. Moreover, it appears that a negative UHI intensity exists during the day and the night (i.e., the temperature in the urban area is lower than the temperature in the rural area), which is also observed in many cities. Several factors could be the cause of negative UHI values. In this case, two of the main reasons could be the hot air that, coming from the desert, increases the air temperature in the rural area more than in the urban area and the presence of the sea breeze [9,10] mitigates the urban overheating when the cold wind blows from the seaside. This confirms the many experimental investigations about the effect of the sea breeze on the magnitude of the UHI [43]. In addition, several studies have demonstrated that the flow of the sea breeze across the city is delayed by the UHI because there is a stagnation zone over the built environments [45,62]; other studies conclude that UHI could accelerate the front of the wind blowing from the seaside and moving across a city, more than in rural areas where it has little or no effect [63,64].

The results of this study, which focuses on wind speed, proved that wind speed has an impact on the magnitude of UHI [58] depending on the prevailing climatic conditions and the wind direction. Similar to the study of Sydney in 2017, a relation between the wind speed and UHI intensity was identified in this study [2]. Moreover, other studies in Chicago, Salamanca, and Granada explain the impact of UHI under ambient climatic conditions such as cloud cover and wind [50,65,66].

Regarding the cloud cover, it was observed that the night-time UHI intensity in the urban and suburban areas reaches the highest values in the presence of clear sky days, due to the larger amount of heat that is absorbed by the built area during the day and then released at night.

In the urban station (i.e., station 1), a reverse correlation between the wind speed and UHI was observed. For most wind directions, when the wind speed increased, the UHI decreased. However, when the wind is coming from the North, the correlation is close to zero. These results are consistent with the findings of previous studies that have examined the impact of wind speed on UHI magnitude in various areas such as Ibadan and Ulaanbaatar [3,47,48].

Moreover, a significantly strong and negative correlation was observed in the suburban area (i.e., station 2) between the levels of the air temperature and wind speed and the magnitude of the UHI. It was found that the effect of UHI intensity in all wind directions is decreasing or even minimized with the increase in the temperature and especially of the wind speed. A possible explanation is that when the wind speed increases, the temperature variations (i.e., between 48.6 °C and 12.3 °C with an average value of about 29.5 °C) gradually decrease since higher wind speeds are connected with the advection of air masses that tend to minimize the temperature differences. This relation has been observed in several cities around the world [51,67], as well as in the urban station (i.e., station 1).

It was observed that when the wind speed is low or zero, the UHI has a positive value because the temperature in the urban area (where the urban surface easily absorbs solar radiation) is higher than in the rural area (where the natural surface has less absorption). When the wind speed is very high, the UHI effect disappears due to the heat dissipation produced by the wind. Similarly, the study of Athens found that the high temperatures resulted from solar radiation when the wind speed was low [52].

The data collected by reference station 3 (i.e., Saih Al Salem), which is located in a rural desert village far from the seacoast, show a moderate to a strong correlation between the UHI and the temperature and wind speed, for all wind directions. This is similar to the relation found in the studies of Ibadan and Sydney [2,47].

In the case of UHI intensity $_{T2-T3}$, the UHI magnitude varies between 11.5 °C and −12.8 °C. It is found that the results for the urban station are similar to those observed in the suburban station through the correlation methodology used for the UHI magnitude and the climate parameters. It was also observed that there were no significant differences between the two stations during the 5 year period of interest. This indicates that when the wind comes from the sea (i.e., western and north-western) directions, there is a moderate negative relation between the UHI and both the temperature and wind speed. The negative results could be due to the additional anthropogenic heat and the cooling rates caused by evaporation. Data analysis showed an inverse relation between the wind speed and the UHI intensity where when the former increased, the latter decreased. In contrast, for all other wind directions, the analysis indicated no correlation between the UHI intensity and both temperature and wind speed. These results are aligned with those of previous studies conducted in Asian and Australian cities [58].

The results show that, in the case of station 3, there is a significant correlation between the meteorological parameters and UHI. A moderate to a quite strong relation is found between this UHI magnitude and the two parameters (i.e., temperature and wind speed) for all the wind directions compared to the previous case of the UHI intensity $_{T1-T3}$. These results could occur with synoptic weather and specific conditions created by the combination of climate parameters, which is consistent with the findings from the previous studies carried out in various regions [2,47].

The results of this study indicate that wind speed and temperature have an impact on the urban heat island depending on the prevailing weather conditions on whether the wind comes from the sea or from the desert. However, the urban area is the one most affected. In fact, the relations, found in this study, between the meteorological parameters and UHI, have several implications for future urban planning. Strategies are needed to reduce the temperatures and have better cooling systems within urban areas in order to control urban heat islands. There is a need to establish guidelines for urban planning and design so that urbanization is sustainable. Worth noting is the fact that one of the challenges faced in this study was the limited availability of meteorological parameters and adequate monitoring data covering a greater number of years.

5. Conclusions

The aim of this paper was to examine the correlation between the intensity of the canopy urban heat island and the various meteorological parameters (e.g., temperature, wind speed, and wind direction) when the wind is coming from different directions such as from the seaside and the desert side. Hourly data of climate parameters were derived from five years (2014–2018) of monitoring from three meteorological stations located in an urban, suburban, and a rural area of Dubai. Six clusters of wind directions were identified to perform a cluster analysis on the available set of climatic data.

It was found that the temperature and wind speed results differed between the western (seaside) and eastern (desert side) parts of Dubai. The UHI intensity varies with the wind direction and meteorological conditions. Regarding the UHI intensity $_{T1-T3}$, it was found that when the wind in the urban area was coming from the desert, the temperature and the UHI were almost independent. However, a relation exists when the wind is coming from the seaside (i.e., sea breeze). Therefore, the wind speed influences the UHI intensity depending on the synoptic climate conditions and wind directions. A reverse correlation was found between the wind speed and UHI intensity for all wind directions, except the North direction where no correlation was found in the urban area. In the suburban area, an inverse relation was observed between the wind speed and the magnitude of UHI, for all wind directions. At the same time, a moderate to strong correlation was found between the UHI and both temperature and wind speed for all wind directions in the rural area.

In contrast, in the case of the UHI intensity $_{T2-T3}$, it was observed that in the urban and suburban areas, there was a similar correlation between the UHI and the climate parameters when the wind was blowing inland from the sea due to additional anthropogenic heat and evaporation. No correlation was observed for all other wind directions. A moderate to a strong association was found between the UHI and the two climate parameters for all the wind directions in the open zone of the desert, where the reference station (i.e., rural area) is located, due to a combination of climate parameters and synoptic conditions.

From that perspective, the wind that flows from the seaside or the desert side is completely different as the air coming from the sea is cooler, which decreases the temperature and minimizes the UHI. Thus, an onshore wind cools the air temperature of the area depending on the wind speed; hence, the higher the wind speed, the lower will be the magnitude of the UHI. On the other hand, the wind coming from the desert is warmer, producing dry hot weather conditions and increasing the temperature. However, occasionally the wind blowing from the desert has different effects on temperature depending on other factors.

This study, investigating the canopy UHI phenomenon in the coastal and desert areas of Dubai, contributes to a deeper understanding of the local microclimate and urban overheating in desert regions that is essential to inform the climate-resilient urban design and planning. It was observed that the sea breeze plays a vital role in the urban zone and the seashore area, contributing to the mitigation of the summer urban and suburban warming. The results of this study confirmed that additional mitigation strategies for heat reduction should be implemented in desert cities to reduce the thermal stress in the urban ecosystem and avoid many issues that can be caused by high urban temperatures such as heat-related illness and mortality and uncomfortable outside areas.

Further studies should be performed to better understand the overall UHI phenomenon in Dubai, for example also investigating the surface UHI and the boundary UHI with the aim to cover different scales of investigation.

Author Contributions: Conceptualization, G.P. and M.S.; methodology, M.S.; formal analysis, A.M.; investigation, A.M., G.P. and M.S.; resources, E.T.; data curation, A.M. and G.P.; writing—original draft preparation, A.M. and G.P.; writing—review and editing, A.M., G.P., E.T. and M.S.; visualization, A.M. and G.P.; supervision, G.P. and M.S. All authors have read and agreed to the published version of the manuscript.

Funding: This research received no external funding.

Acknowledgments: The authors express their gratefulness to the UAE National Center of Meteorology (NCM) for providing the data used in this study.

Conflicts of Interest: The authors declare no conflict of interest.

References

1. Un.org. World's Population Increasingly Urban with More Than Half Living in Urban Areas. Available online: https://www.un.org/en/development/desa/news/population/world-urbanization-prospects-2014.html#:~{}:text=The%202014%20revision%20of%20the,population%20between%202014%20and%202050 (accessed on 6 June 2020).
2. Santamouris, M.; Haddad, S.; Fiorito, F.; Osmond, P.; Ding, L.; Prasad, D.; Zhai, X.; Wang, R. Urban Heat Island and Overheating Characteristics in Sydney, Australia. An Analysis of Multiyear Measurements. *Sustainability* **2017**, *9*, 712. [CrossRef]

3. Oke, T.R. The energetic basis of the urban heat island. *Q. J. R. Meteorol. Soc.* **1982**, *108*, 1–24. [CrossRef]
4. Oke, T.R.; Johnson, G.T.; Steyn, D.G.; Watson, I.D. Simulation of surface urban heat islands under 'ideal' conditions at night part 2: Diagnosis of causation. *Bound. Layer Meteorol.* **1991**, *56*, 339–358. [CrossRef]
5. Santamouris, M. Recent progress on urban overheating and heat island research. Integrated assessment of the energy, environmental, vulnerability and health impact. Synergies with the global climate change. *Energy Build.* **2020**, *207*, 109482. [CrossRef]
6. Santamouris, M. Cooling the cities—A review of reflective and green roof mitigation technologies to fight heat island and improve comfort in urban environments. *Sol. Energy* **2014**, *103*, 682–703. [CrossRef]
7. Santamouris, M.; Paolini, R.; Haddad, S.; Synnefa, A.; Garshasbi, S.; Hatvani-Kovacs, G.; Gobakis, K.; Yenneti, K.; Vasilakopoulou, K.; Feng, J.; et al. Heat mitigation technologies can improve sustainability in cities. An holistic experimental and numerical impact assessment of urban overheating and related heat mitigation strategies on energy consumption, indoor comfort, vulnerability and heat-related mortality and morbidity in cities. *Energy Build.* **2020**, *217*, 110002. [CrossRef]
8. Santamouris, M.; Kolokotsa, D. *Urban Climate Mitigation Techniques*; Routledge: London, UK, 2016.
9. Sasaki, Y.; Matsuo, K.; Yokoyama, M.; Sasaki, M.; Tanaka, T.; Sadohara, S. Sea breeze effect mapping for mitigating summer urban warming: For making urban environmental climate map of Yokohama and its surrounding area. *Urban Clim.* **2018**, *24*, 529–550. [CrossRef]
10. Kawamoto, Y.; Yoshikado, H.; Ooka, R.; Hayami, H.; Huang, H.; Khiem, M.V. Sea Breeze Blowing into Urban Areas: Mitigation of the Urban Heat Island Phenomenon. In *Ventilating Cities: Air-flow Criteria for Healthy and Comfortable Urban Living*; Springer: Dordrecht, The Netherlands, 2012; pp. 11–32. [CrossRef]
11. Comstock, M.; Garrigan, C.; Pouffary, S.; Feraudy, T.d.; Halcomb, J.; Hartke, J.J.U.N.E.P. *Building Design and Construction: Forging Resource Efficiency and Sustainable Development*; United National Environmental Program (UNEP): Nairobi, Kenya, 2012; pp. 1–24.
12. Santamouris, M. *Minimizing Energy Consumption, Energy Poverty and Global and Local Climate Change in the Built Environment: Innovating to Zero: Causalities and Impacts in a Zero Concept World*; Elsevier: Amsterdam, The Netherlands, 2019.
13. Agency, I.E.; Birol, F. *World Energy Outlook 2013*; International Energy Agency Paris: Paris, France, 2013.
14. Wang, Y.; Du, H.; Xu, Y.; Lu, D.; Wang, X.; Guo, Z. Temporal and spatial variation relationship and influence factors on surface urban heat island and ozone pollution in the Yangtze River Delta, China. *Sci. Total Environ.* **2018**, *631–632*, 921–933. [CrossRef]
15. Santamouris, M. *Energy and Climate in the Urban Built Environment*; CRC Press LLC: London, UK, 2001. [CrossRef]
16. Santamouris, M.; Cartalis, C.; Synnefa, A.; Kolokotsa, D. On the impact of urban heat island and global warming on the power demand and electricity consumption of buildings—A review. *Energy Build.* **2015**, *98*, 119–124. [CrossRef]
17. Santamouris, M. Cooling the buildings—past, present and future. *Energy Build.* **2016**, *128*, 617–638. [CrossRef]
18. Kovats, R.S.; Hajat, S.; Wilkinson, P. Contrasting patterns of mortality and hospital admissions during hot weather and heat waves in Greater London, UK. *Occup. Environ. Med.* **2004**, *61*, 893–898. [CrossRef]
19. Baccini, M.; Biggeri, A.; Accetta, G.; Kosatsky, T.; Katsouyanni, K.; Analitis, A.; Anderson, H.; Bisanti, L.; D'Ippoliti, D.; Danova, J.; et al. Heat Effects on Mortality in 15 European Cities. *Epidemiology* **2008**, *19*, 711–719. [CrossRef] [PubMed]
20. Goggins, W.; Chan, E.Y.; Ng, E.; Ren, C.; Chen, L. Effect Modification of the Association between Short-term Meteorological Factors and Mortality by Urban Heat Islands in Hong Kong. *PLoS ONE* **2012**, *7*, e38551. [CrossRef] [PubMed]
21. Mirzaei, P.; Haghighat, F. Approaches to study Urban Heat Island—Abilities and limitations. *Build. Environ.* **2010**, *45*, 2192–2201. [CrossRef]
22. Kato, S.; Yamaguchi, Y. Estimation of storage heat flux in an urban area using ASTER data. *Remote Sens. Environ.* **2007**, *110*, 1–17. [CrossRef]
23. Zhang, X.; Zhong, T.; Feng, X.; Wang, K. Estimation of the relationship between vegetation patches and urban land surface temperature with remote sensing. *Int. J. Remote Sens.* **2009**, *30*, 2105–2118. [CrossRef]
24. epa.gov. Heat Islands. Available online: https://www.epa.gov/heat-islands/heat-island-compendium (accessed on 12 June 2020).

25. Lazzarini, M.; Molini, A.; Marpu, P.; Ouarda, T.M.J.; Ghedira, H. Urban climate modifications in hot desert cities: The role of land cover, local climate, and seasonality. *Geophys. Res. Lett.* **2015**, *42*, 9980–9989. [CrossRef]
26. Lazzarini, M.; Marpu, P.; Ghedira, H. Temperature-land cover interactions: The inversion of urban heat island phenomenon in desert city areas. *Remote Sens. Environ.* **2013**, *130*, 136–152. [CrossRef]
27. Charabi, Y.; Bakhit, A. Assessment of the canopy urban heat island of a coastal arid tropical city: The case of Muscat, Oman. *Atmos. Res.* **2011**, *101*, 215–227. [CrossRef]
28. Radhi, H.; Fikry, F.; Sharples, S. Impacts of urbanisation on the thermal behaviour of new built up environments: A scoping study of the urban heat island in Bahrain. *Landsc. Urban Plan.* **2013**, *113*, 47–61. [CrossRef]
29. Al-Sallal, K.; Al-Rais, L. Outdoor airflow analysis and potential for passive cooling in the modern urban context of Dubai. *Renew. Energy* **2012**, *38*, 40–49. [CrossRef]
30. u.ae/en. about-the-uae/the-seven-emirates/dubai. Available online: https://u.ae/en/about-the-uae/the-seven-emirates/dubai (accessed on 23 May 2020).
31. dsc.gov.ae. Themes > Population and Vital Statistics. Available online: https://www.dsc.gov.ae/en-us/Themes/Pages/Population-and-Vital-Statistics.aspx?Theme=42&year=2014#DSC_Tab1 (accessed on 20 March 2020).
32. weatherspark.com. Average Weather at Dubai International Airport United Arab Emirates. Available online: https://weatherspark.com/y/148889/Average-Weather-at-Dubai-International-Airport-United-Arab-Emirates-Year-Round (accessed on 20 March 2020).
33. ncm.ae. Radar UAE. Available online: https://www.ncm.ae/en#!/Radar_UAE_Merge/26 (accessed on 23 March 2020).
34. dubaidxbairport.com. Dubai International Airport. Available online: https://www.dubaidxbairport.com/ (accessed on 27 March 2020).
35. cnbc.com. Dubai International Airport Installs 15,000 Solar Panels. Available online: https://www.cnbc.com/2019/07/17/dubai-international-airport-installs-15000-solar-panels.html (accessed on 27 March 2020).
36. fscloudport.com. Al Maktoum International Airport (OMDW). Available online: http://www.fscloudport.com/atk/fscp.nsf/c9482105febd1beb802583bb006e932b/5f8ee3f40129ce9b802583bb006ec4f7?OpenDocument (accessed on 1 April 2020).
37. dsc.gov.ae. Population Bulletin Emirate of Dubai 2018. Available online: https://www.dsc.gov.ae/Publication/Population%20Bulletin%20Emirate%20of%20Dubai%202018.pdf (accessed on 3 April 2020).
38. thenational.ae. Dubai Ruler launches Marmoom Desert Conservation Reserve. Available online: https://www.thenational.ae/uae/environment/dubai-ruler-launches-marmoom-desert-conservation-reserve-1.696015 (accessed on 3 April 2020).
39. citypopulation.de. UAE: Division of Dubai. Available online: https://www.citypopulation.de/en/uae/dubai/admin/ (accessed on 10 June 2020).
40. Camilloni, I.; Barrucand, M. Temporal variability of the Buenos Aires, Argentina, urban heat island. *Theor. Appl. Climatol.* **2011**, *107*, 47–58. [CrossRef]
41. von Glasow, R.; Jickells, T.; Baklanov, A.; Carmichael, G.; Church, T.; Gallardo, L.; Hughes, C.; Kanakidou, M.; Liss, P.; Mee, L.; et al. Megacities and Large Urban Agglomerations in the Coastal Zone: Interactions Between Atmosphere, Land, and Marine Ecosystems. *AMBIO* **2012**, *42*, 13–28. [CrossRef] [PubMed]
42. Yoshikado, H. Numerical Study of the Daytime Urban Effect and Its Interaction with the Sea Breeze. *J. Appl. Meteorol. (1988–2005)* **1992**, *31*, 1146–1164.
43. Dandou, A.; Tombrou, M.; Soulakellis, N. The Influence of the City of Athens on the Evolution of the Sea-Breeze Front. *Bound. Layer Meteorol.* **2008**, *131*, 35–51. [CrossRef]
44. Santamouris, M.; Papanikolaou, N.; Livada, I.; Koronakis, I.; Georgakis, C.; Argiriou, A.; Assimakopoulos, D.N. On the impact of urban climate on the energy consumption of buildings. *Sol. Energy* **2001**, *70*, 201–216. [CrossRef]
45. Sakaida, K.; Egoshi, A.; Kuramochi, M. Effects of Sea Breezes on Mitigating Urban Heat Island Phenomenon: Vertical Observation Results in the Urban Center of Sendai. *Jpn. Prog. Climatol.* **2011**, 11–16.
46. Erell, E.; Williamson, T. Intra-urban differences in canopy layer air temperature at a mid-latitude city. *Int. J. Climatol.* **2007**, *27*, 1243–1255. [CrossRef]
47. Anibaba, B.W.; Durowoju, O.S.; Adedeji, O.I. Assessing the Significance of Meteorological Parameters to the Magnitude of Urban Heat Island (Uhi). *Ann. Univ. Oradea, Geogr. Ser.* **2019**, *29*, 30–39. [CrossRef]

48. Ganbat, G.; Han, J.-Y.; Ryu, Y.-H.; Baik, J.-J. Characteristics of the urban heat island in a high-altitude metropolitan city, Ulaanbaatar, Mongolia. *Asia-Pac. J. Atmos. Sci.* **2013**, *49*, 535–541. [CrossRef]
49. Li, D.; Sun, T.; Liu, M.; Wang, L.; Gao, Z. Changes in Wind Speed under Heat Waves Enhance Urban Heat Islands in the Beijing Metropolitan Area. *J. Appl. Meteorol. Climatol.* **2016**, *55*, 2369–2375. [CrossRef]
50. Alonso, M.S.; Fidalgo, M.R.; Labajo, J.L. The urban heat island in Salamanca (Spain) and its relationship to meteorological parameters. *Clim. Res.* **2007**, *34*, 39–46. [CrossRef]
51. Sundborg, Å. Local Climatological Studies of the Temperature Conditions in an Urban Area. *Tellus* **1950**, *2*, 222–232. [CrossRef]
52. Papanikolaou, N.M.; Livada, I.; Santamouris, M.; Niachou, K. The Influence of Wind Speed on Heat Island Phenomena in Athens, Greece. *Int. J. Vent.* **2008**, *6*, 337–348. [CrossRef]
53. Sheng, L.; Tang, X.; You, H.; Gu, Q.; Hu, H. Comparison of the urban heat island intensity quantified by using air temperature and Landsat land surface temperature in Hangzhou, China. *Ecol. Indic.* **2017**, *72*, 738–746. [CrossRef]
54. Santamouris, M. On the energy impact of urban heat island and global warming on buildings. *Energy Build.* **2014**, *82*, 100–113. [CrossRef]
55. Liu, W.; Ji, C.; Zhong, J.; Jiang, X.; Zheng, Z. Temporal characteristics of the Beijing urban heat island. *Theor. Appl. Climatol.* **2006**, *87*, 213–221. [CrossRef]
56. Skoulika, F.; Santamouris, M.; Kolokotsa, D.; Boemi, N. On the thermal characteristics and the mitigation potential of a medium size urban park in Athens, Greece. *Landsc. Urban Plan.* **2014**, *123*, 73–86. [CrossRef]
57. Mihalakakou, G.; Flocas, H.A.; Santamouris, M.; Helmis, C.G. Application of Neural Networks to the Simulation of the Heat Island over Athens, Greece, Using Synoptic Types as a Predictor. *J. Appl. Meteorol. (1988–2005)* **2002**, *41*, 519–527. [CrossRef]
58. Santamouris, M. Analyzing the heat island magnitude and characteristics in one hundred Asian and Australian cities and regions. *Sci. Total Environ.* **2015**, *512–513*, 582–598. [CrossRef]
59. Schatz, J.; Kucharik, C. Urban climate effects on extreme temperatures in Madison, Wisconsin, USA. *Environ. Res. Lett.* **2015**, *10*, 094024. [CrossRef]
60. Lee, S.-H.; Baik, J.-J. Statistical and dynamical characteristics of the urban heat island intensity in Seoul. *Theor. Appl. Climatol.* **2010**, *100*, 227–237. [CrossRef]
61. Kim, Y.-H.; Baik, J.-J. Maximum Urban Heat Island Intensity in Seoul. *J. Appl. Meteorol. (1988–2005)* **2002**, *41*, 651–659. [CrossRef]
62. Yoshikado, H.; Tsuchida, M. High Levels of Winter Air Pollution under the Influence of the Urban Heat Island along the Shore of Tokyo Bay. *J. Appl. Meteorol.* **1996**, *35*, 1804–1813. [CrossRef]
63. Khan, S.; Simpson, R. Effect Of A Heat Island On The Meteorology Of A Complex Urban Airshed. *Boun. Layer Meteorol.* **2001**, *100*, 487–506. [CrossRef]
64. Freitas, E.; Rozoff, C.; Cotton, W.; Dias, P.S. Interactions of an urban heat island and sea-breeze circulations during winter over the metropolitan area of São Paulo, Brazil. *Bound. Layer Meteorol.* **2006**, *122*, 43–65. [CrossRef]
65. Ackerman, B. Temporal March of the Chicago Heat Island. *J. Clim. Appl.Meteorol.* **1985**, *24*, 547–554. [CrossRef]
66. Montávez, J.P.; Rodríguez, A.; Jiménez, J.I. A study of the Urban Heat Island of Granada. *J. R. Meteorol. Soc.* **2000**, *20*, 899–911. [CrossRef]
67. Camilloni, I.s.; Barros, V. On the Urban Heat Island Effect Dependence on Temperature Trends. *Clim. Chang.* **1997**, *37*, 665–681. [CrossRef]

Article

On the Efficiency of Using Transpiration Cooling to Mitigate Urban Heat

Kai Gao [1,*], Mattheos Santamouris [1,2] and Jie Feng [1]

[1] Faculty of Built Environment, University of New South Wales, Sydney 2052, Australia; m.santamouris@unsw.edu.au (M.S.); jie.feng@unsw.edu.au (J.F.)
[2] Anita Lawrence Chair on High Performance Architecture, Faculty of Built Environment, University of New South Wales, Sydney 2052, Australia
* Correspondence: kai.gao@unsw.edu.au; Tel.: +61-452098190

Received: 19 April 2020; Accepted: 29 May 2020; Published: 1 June 2020

Abstract: Trees are considered to be effective for the mitigation of urban overheating, and the cooling capacity of trees mainly comes from two mechanisms: transpiration and shading. This study explores the transpiration cooling of large trees in urban environments where the sea breeze dominates the climate. In the experiment, sap flow sensors were used to measure the transpiration rate of two large trees located in Sydney over one year. Also, the temperature difference between the inside and outside of the canopy, as well as the vertical temperature distribution below the canopy, were measured during summer. In this experiment, the temperature under the canopies decreased by about 0.5 degrees from a 0.5 m height to a 3.5 m height, and the maximum temperature difference between the inside and outside of the canopy was about 2 degrees. After applying a principal component analysis of multiple variables, we found that when a strong sea breeze is the primary cooling mechanism, the sap flow still makes a considerable contribution to cooling. Further, the sea breeze and the transpiration cooling of trees are complementary. In conclusion, the characteristics of synoptic conditions must be fully considered when planting trees for mitigation purposes. Since the patterns of sea breeze and sap often do not coincide, the transpiration cooling of trees is still effective when the area is dominated by sea breeze.

Keywords: transpiration cooling; coastal cities; sap flow

---

1. Introduction

Climate change and the urban heat island effect have caused a worldwide issue of urban overheating [1]. Urban overheating will increase the total energy consumption and peak electricity demands, thereby deteriorating outdoor and indoor thermal comfort, as well as the health of the public [2–4]. In order to mitigate urban overheating, researchers have proposed a variety of options: integrating cool materials, such as reflective materials [5] and radiative cooling surfaces, into facades and roofs [6]; using cool pavement on roads [4,7]; and expanding green spaces in the urban environment [8].

In the past few decades, the mitigation effect of trees has been proven effective by a large number of studies [9–11]. The difference between the ambient temperature and the temperature in the canopy is usually used as an essential indicator of plant cooling capacity. For instance, the daily average difference of the air temperature between the shady area of a street tree and an open space is 0.9 °C in Melbourne, Australia [12]; 1 °C in Munich, Germany [13]; and up to 2.8 °C in Southeast Brazil [14]. The cooling effect of trees mainly comes from shading and transpiration. The cooling effect caused by shading at a macroscopical level is very limited, as the albedo of the tree canopy is very similar to that of the ground surface, indicating that the total energy balance is not significantly affected by the presence of trees. Transpiration, on the other hand, alters the energy balance of the whole area by converting sensible heat flux to latent heat flux and has a cooling effect both locally and on a large scale.

When sufficient water is supplied, the transpiration rate can be maximized, providing a considerable cooling effect. Plenty of studies have shown that the air temperature over vegetated land is significantly reduced after applying irrigation [15–17].

Since research in forestry and agronomy usually focuses on large-scale crops and woodland, the heat flux over the land is easier to quantify to determine the transpiration cooling effect. However, in cities, where the surface is highly heterogeneous, the heat flux method is not suitable. Existing studies on urban tree cooling often use the sap flow method to measure the transpiration rate of a single tree or a small number of trees.

When switching to microscale studies, the transpiration cooling effect of urban trees can be explored more in depth. Researchers have found that the cooling effect of trees is closely related to their species [18–20] and hydraulic architecture. The leaf area index (LAI) is considered to be a particularly critical parameter that determines the cooling capacity of trees. Trees with a high LAI can achieve more efficient shading and exhibit higher transpiration potential.

However, microscale studies also face difficulties. The temperature difference indicator reflects the cooling capacity of a plant to a certain extent, but this indicator can be environmentally sensitive. Remoting sensing research [21] has found that in the core areas of several U.S. cities, the urban tree cooling effect was significantly affected by the surrounding thermal conditions. Compared with studies in forestry, experiments conducted in cities usually found larger temperature decreases under the canopy. For instance, an experiment in the Munich urban area [22] reported that under the canopy, temperature reduced by 3.5 degrees, from 0.5 m above the ground to 4 m above the ground. However, in a *Picea abies* L. forest, research [23] found that the temperature only reduces by one degree every 4 m vertically under the canopy. Possibly, the warming effect of the paved street prevented the downward penetration of the tree cooling effect in the experiment conducted in Munich [22]. This possibility emphasizes the influence of the energy balance of the local environment on the magnitude of the vertical temperature difference. Moreover, since a tree cooling mechanism differs from that of short vegetation, the vertical temperature cannot be used as the only indicator for transpirational cooling. Short vegetation directly provides cooling on the ground, while the transpiration cooling effect of trees needs to penetrate several meters to reach the ground from the canopy. The overall cooling performance is determined only after determining both the vertical temperature difference and the horizontal temperature difference between the inside and outside of the canopy.

At the same time, weather conditions could also affect the cooling effects of trees [24]. The research in [21] observed the variation of the urban tree cooling effect under different climatic conditions, and in [25], more effective cooling was observed from trees in the dessert compared to trees on the coast. In scenarios where strong advection exists, the temperature difference under the canopy in the vertical direction and that between the inside and outside of the canopy may be affected. On the one hand, if the wind is dry and hot, the vapour pressure deficit (VPD) in the canopy will increase, and transpiration cooling will be stimulated [26]. However, continuous dry and hot wind will also blow away the cold air that is mass-produced by the trees [27]. On the other hand, if the wind is cold and humid, such as a sea breeze, transpiration can be weakened. At the same time, the canopy can directly block the wind, making it receive less cold from the wind compared to that received by the outside. If the space outside the canopy is also shaded, the temperature under the canopy can be even hotter than that outside. This possible phenomenon indicates that the effect of the wind may completely offset the cooling effect of transpiration. However, the effect of sea breeze on the cooling of trees has not been sufficiently studied, and many issues remain unclear. For example, in cities where the sea breeze is the dominant cooling mechanism, does planting more trees have an obvious cooling effect? How much tree cooling contributes to local temperature distribution? Answering such questions is crucial for mitigating overheating in coastal cities.

At the same time, some cities are affected by both desert and marine climates, such as Sydney, Australia. In Sydney, a synoptic condition exists where the east side of Sydney is adjacent to the Pacific Ocean and is affected by strong humid and cold sea breeze, while the west side of Sydney faces a vast desert and is influenced by the dry and hot desert winds. If a large number of trees are planted to alleviate urban overheating in the city, how should we arrange these trees spatially to achieve the most effective cooling? An analysis of the impact of sea breeze on tree cooling can help answer this question.

In this paper, when discussing the effect of the transpiration cooling of big trees, weather station data are used to include the impact of Sydney's synoptic conditions, and the temperature difference between the inside and outside of the canopy was measured. The experimental site in this paper is about 2 km away from the coastline and is significantly affected by the sea breeze. As the elevation of the experimental site is higher than that of the eastern coast, the sea breeze can easily reach this area. Thus trees here are significantly affected by the sea breeze. After obtaining the transpiration rate, temperature difference, and wind speed, a principal component analysis (PCA) was used to determine the contribution of transpiration to the temperature difference inside and outside the canopy. At the same time, this research also studies the factors that affect the transpiration rates of plants. The results show that the sea breeze does play a crucial role in ooling, and sap flow also makes a considerable contribution to cooling. The analysis also shows that the daily peaks of sea breeze and sap flow do not overlap, indicating that the introduction of trees in coastal cities can complement the cooling of the sea breeze. This result answers the above questions and affirms the plant's cooling potential under sea breeze conditions.

In the study, attention was also paid to the vertical temperature gradient below the tree, and the influence of the ground surface on the vertical temperature distribution was discussed according to the measurement results. This result further emphasizes the effect of the ground surface on the cooling capacity of trees and highlights that when measuring the cooling effect of trees in a microenvironment, the vertical temperature difference cannot be used as the only criterion to assess the cooling capacity of plants; the assessments should instead be made after correcting the possible impact of the surface. We believe that this study makes an important contribution to rationalizing the layout of urban greenery to achieve a better mitigation effect.

2. Methods

2.1. Experiment Design

In order to analyse the transpiration cooling effects of large urban trees, the transpiration rate and surrounding temperature of two large Melaleuca trees were measured. The transpiration rate was measured using sap flow sensors. The studied site was located on campus at the University of New South Wales in Sydney.

The recording of sap flow and soil moisture began in early 2018 and continued until May 2019. To obtain an understanding of the trees' cooling effect, multiple temperature sensors were placed under the canopies and outside the canopies of Maleluca from January 2019 to March 2019. The temperature sensors were placed outside canopy. Since aluminum foil was also proven to be an effective solar protection measure in [28], we used it to shield the sensors from direct solar radiation. At the same time, the vertical air temperature distribution under the canopies was recorded for 2 weeks in the middle of February 2019. The weather data were extracted from the nearest weather station.

2.2. Measurement Set Up

2.2.1. Site Details

Two almost-identical Melaleuca trees were investigated, as shown in Figure 1. The lawn at the university of New South Wales (UNSW) campus has dense grass and is regularly irrigated. Hence, the soil in the studied site has remained moist for the entire period of measurement, ensuring that the transpiration of trees is not limited by soil water content.

(a) site plan of the researched area

(b) weather station information

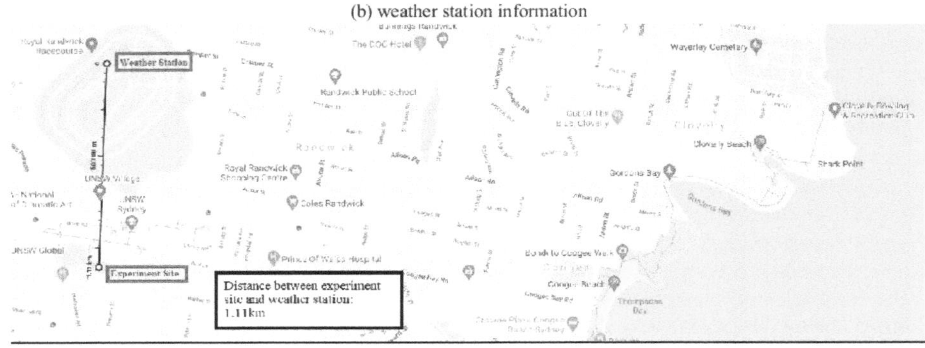

Figure 1. (**a**) site plan of the studied area; the red circles are the canopies of the researched trees. (**b**) weather station information.

The essential information of the studied trees is measured and listed in Table 1. The canopy of the trees starts at around 3 m of height. Site weather data were extracted from the nearest weather station (Randwick weather station, 33.92° S, 151.2° E), which is in similar condition to the experimental site. The vapour pressure deficit used in our experiment was calculated from the vapour pressure and saturated vapour pressure of the weather station observations. The air temperature, wind direction, and speed were also extracted from the station observation. The site soil surface moisture and soil surface temperature was monitored at 5 cm beneath the soil surface using an Edaphic TEROS-11 model sensor.

Table 1. Tree parameters.

ID	Height	Trunk Diameter at 1 m Height
1	14 m	1 m
2	12 m	0.9 m

2.2.2. Sap Flow Method

In this study, the transpiration rate of trees was reflected by the sap flow rate, and we used Edaphic SF3 model heat pulse sensors to record the sap flow rate. The sap flow rate was recorded in a 30 min time interval, which provided a balance between battery life and experimental accuracy. A sensor was installed on the north side of each tree.

The sensor used analog signals to record the time, and the sensor received the heat pulse signal after the heat pulse was generated from the heat source. Both the sap flow data and soil moisture were recorded using a CR100 datalogger from Campbell Science.

In order to obtain more accurate sap flow data, two methods were used to calculate the sap flow velocity. The Marshall heat pulse method [29] was applied to calculate the slow sap flow rate, while the T-max method [30] was used to calculate the high sap flow rate. The T-max method is written as

$$V_h = \frac{\sqrt{x^2 - 4kt_m}}{t_m}, \tag{1}$$

where V_h is the heat pulse velocity(cm/hr), x is the distance to the axial temp, k is the thermal diffusivity, and t_m is the time it takes the sensor to reach its max temperature after a heat pulse is generated by a heater.

To calculate the slow sap flow rate, the heat pulse method was used, which is written as

$$V_h = \frac{k}{x}\ln\left(\frac{\Delta T_d}{\Delta T_u}\right), \tag{2}$$

where ΔT_d and ΔT_u are the temperature changes recorded by the upstream sensors and downstream sensors, respectively. Later, the heat pulse velocity was transformed into sap flow velocity. Since we were unable to obtain a trunk sample from the studied trees, the calculated sap flow rate was not calibrated. Therefore, the analysis of sap flows only focused on patterns.

2.2.3. Temperature Measurement

The Log tag sensors were used to record the air temperature. Two Log tag temperature sensors were set up and fixed at around 1.5 m height under the canopies; they were attached to the trunk on the south side to avoid direct sunlight. Two other Log tag sensors were placed outside the canopy at the same height as the sensors inside canopy. All Log tags were shaded from direct sunlight, and the measurement interval was 15 min.

The vertical temperature was measured using globe thermal meters. The measurements were conducted under the canopy for two weeks in February 2019, from 12:30 to 16:20 every day, with a measurement interval of 10 min. The height from 0.5 m to 3.5 m was measured with a space interval of 0.3 m, and the vertical temperature distribution was measured manually. The globe thermometers were fixed on supports (tripods). Each time after adjusting the heights of the globe thermometers, the researchers walked away to avoid interference of the human body in the measurements. The readings were recorded after they became stable. To reduce random errors, the measurement of the temperature at each height was repeated 8 times from eight different points, which were all about 1 m away from the trunk in different directions. The average value of the 8 measurements was considered to be the value of the temperature at the specific height.

All sensors were calibrated, and the error of the temperature sensors was negligible (±0.2 °C).

3. Results

3.1. Site Analysis

In Sydney, summer weather is dominated alternately by the wet and cold sea breeze from the east side and the hot and dry desert wind from the west side. Desert wind brings heat, but, at the same time, it facilitates evapotranspiration and improves the cooling effect of evapotranspiration. In contrast, the wet and cold sea breeze has a cooling effect but can also hinder evapotranspiration. The extra moisture produced by the sea breeze can also increase evapotranspiration resistance. In our case, the effect of the sea breeze was substantial because the study area was very close to the coastline. Figure 2 shows an analysis of the wind direction from an observation of the nearest weather station. It can be seen in Figure 2d that most of the wind came from the southeast starting at noon every day during summer 2019.

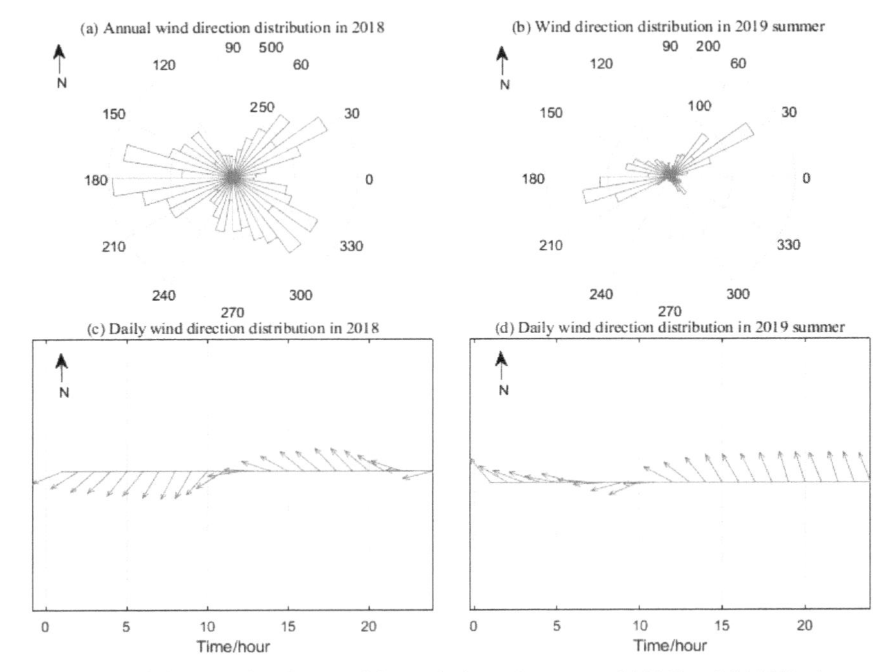

Figure 2. Wind direction distribution of the studied area in summer (**a**) 2018 and (**b**) 2019. Average daily wind direction distribution in summer (**c**) 2018 and (**d**) 2019.

The wind speed of the sea breeze gradually increased in the afternoon, as shown in Figure 3. The wind speed reached its maximum between 17:00 and 19:00 in the summer.

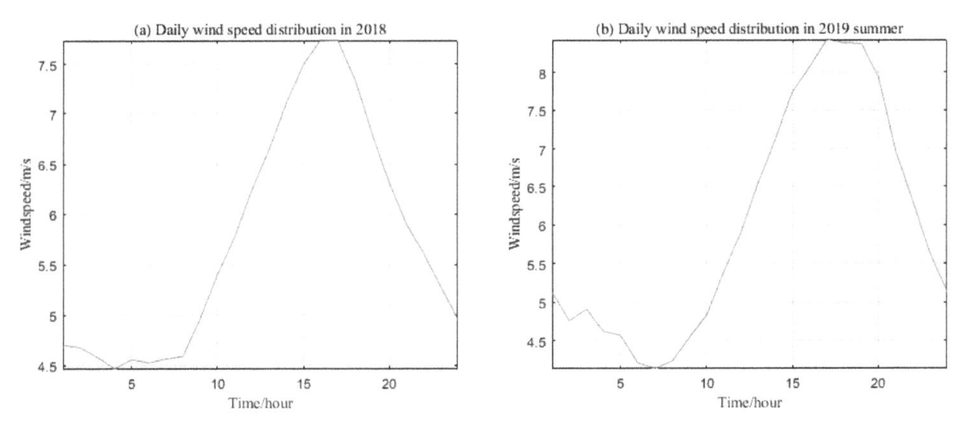

Figure 3. Daily average wind speed variation of the studied area in summer (**a**) 2018 (**b**) 2019.

3.2. Diurnal Sap Flow Patterns

The Diurnal period in this study is defined as the time between 6:00 and 20:00, and we analysed the average sap flow during this period. The canopy temperature is the average temperature recorded by two log tag sensors under the canopies.

As mentioned above, the UNSW campus is regularly irrigated, and the measured Melaleuca canopy is a large tree with a deep root system. Thus, the pattern of sap flow is not limited by the soil moisture content. The analysis shows that the correlation coefficient between the sap flow rate and soil moisture was smaller than 5%. Moreover, the multiple linear regression analysis shows that the diurnal sap flow patterns are positively correlated to both the temperature and the VPD, as shown in Figure 4.

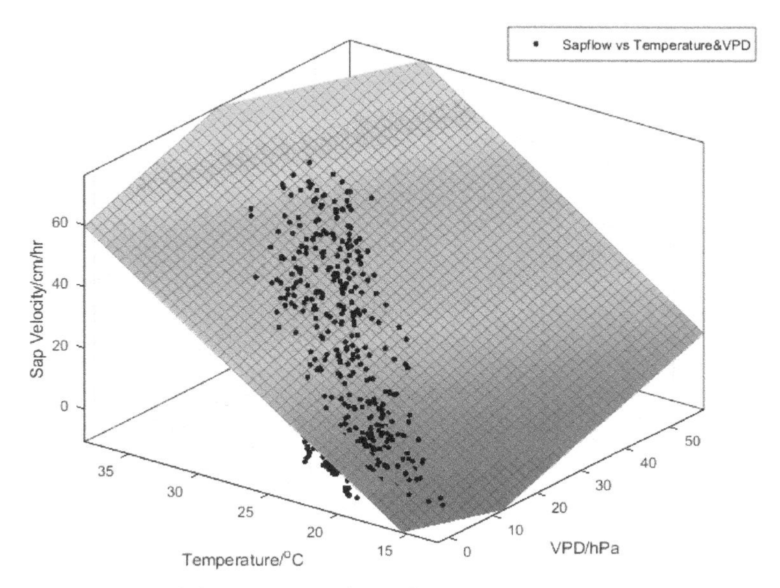

Figure 4. Multilinear regression for sap flow, canopy temperature, and VPD.

A non-linear model, expressed as $f(x) = c - a \times \exp(-b \times x)$, was used to fit the relationship between the VPD and diurnal sap flow, as shown in Figure 5. When considering 95% confidence bounds, the fit results are a = 61.27 (58.54, 63.99) and b = 0.05848 (0.0534, 0.06357), while c is determined as 58.4534. At the same time, a linear regression relationship was built between the sap flow and the canopy temperature, which has an R^2 of 0.65, as shown in Figure 6.

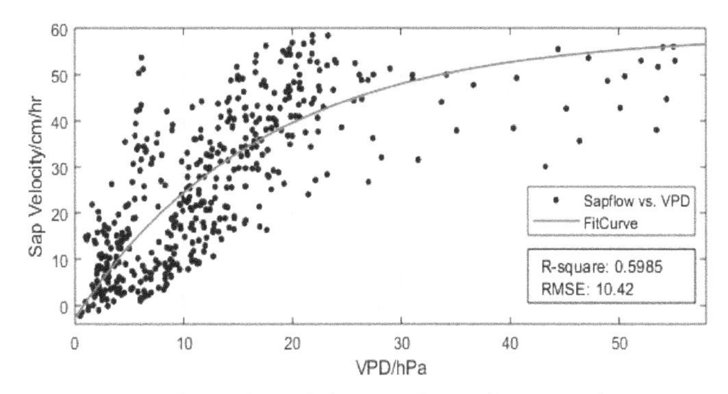

Figure 5. The non-linear fit between the sap flow rate and VPD.

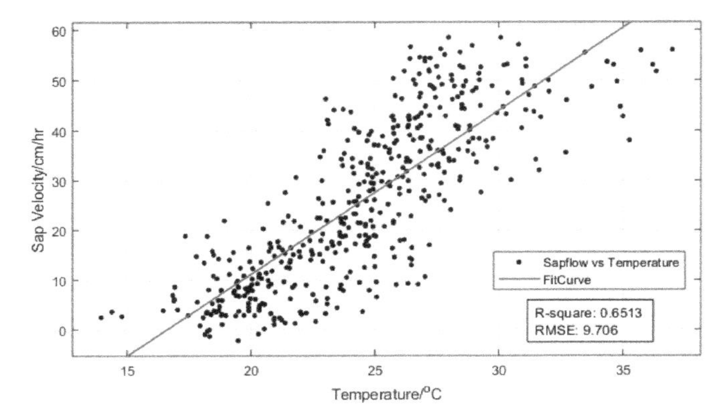

Figure 6. The linear fit between the sap flow rate and canopy temperature.

The Pearson correlation coefficient [31] was also calculated. The correlation coefficient between the diurnal sap flow and diurnal canopy temperature is 0.8, while the correlation coefficient between the diurnal sap flow and diurnal VPD is 0.69.

3.3. *Air Temperature Distribution*

3.3.1. Vertical Temperature Distribution

Measurement of the vertical temperature distribution was performed between 12:30 and 16:30 since the highest daily radiation and temperatures usually occur within these hours. The measurement results show that the maximum temperature difference vertically was only about 0.5 °C. Figure 7 shows the vertical temperature change recorded during the experiment.

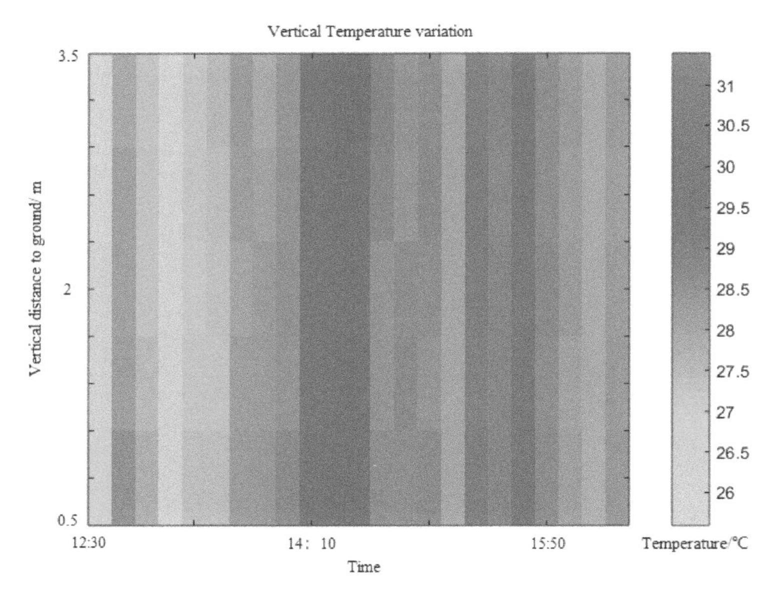

Figure 7. 0.5–3.5 m height vertical temperature distribution under the canopy.

3.3.2. Temperature Difference between the Inside and outside of the Canopy

A linear regression of the temperature difference between the inside and outside of the canopy showed that a stronger sap rate was associated with a larger temperature difference between the inside and outside, as shown in Figure 8a. However, the same relationship was also found in the analysis of VPD and wind speed, as shown in Figure 8b,c.

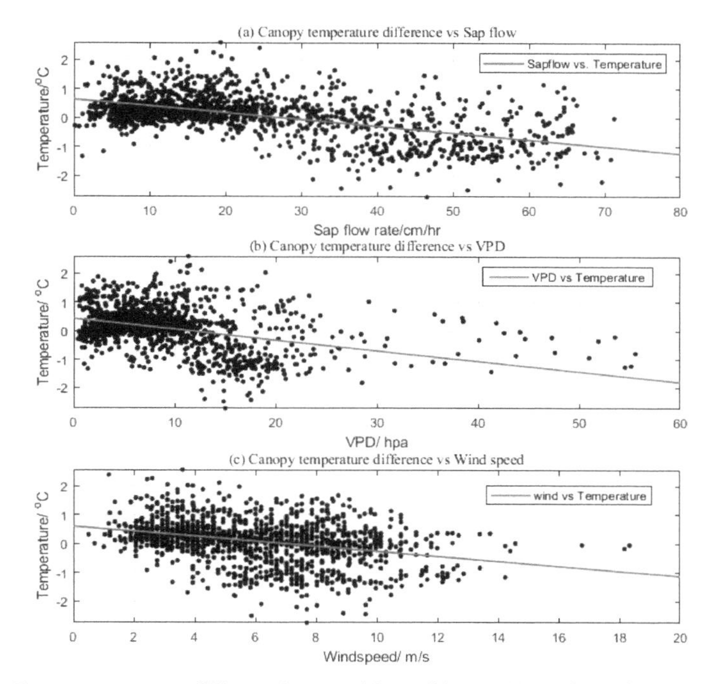

Figure 8. Canopy temperature difference between (**a**) sap; (**b**) VPD; (**c**) wind speed. The R squares are 0.30, 0.15, and 0.14, respectively.

At the same time, most of the variables in this study are highly correlated with each other, especially when strong wind conditions are considered. The variation of the temperature difference inside and outside the canopy can be explained only when the correlation between the observed variables is eliminated. Therefore, a PCA analysis was performed using the VPD, sap flow rate, and wind speed as input observational groups. Ultimately, three groups of components were created. The results showed that principal components 1 and 2 account for 92.04% of the variability in the data set. As shown in Figure 9, in the configuration of principal components 1 and 2, the contributions of wind speed and sap flow rate dominate with similar contributions. The details of the principle components are shown in Table 2.

A multiple linear regression with canopy temperature difference (CTD) was also performed using principal components 1 and 2, and the results are shown in Figure 10. The R square is 0.32, and principal component 1 shows a negative correlation with temperature difference.

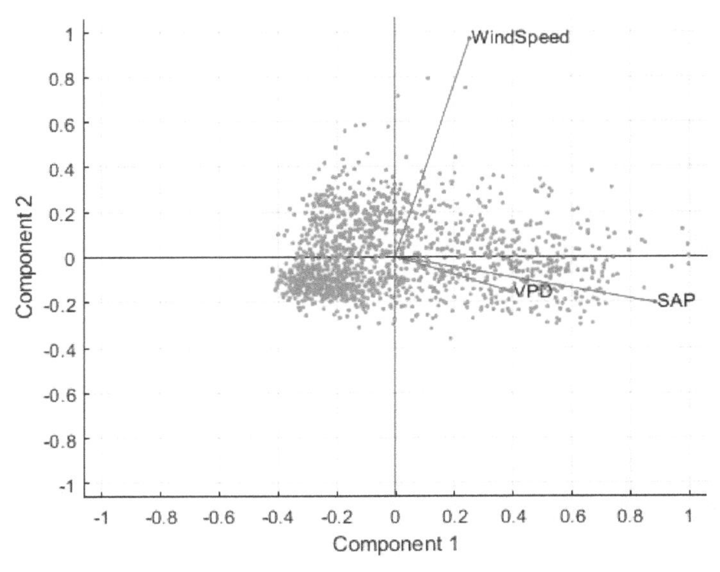

Figure 9. Recentered component configuration after applying PCA to the observations. The lengths of the vectors indicate the contribution of the specific observation.

Table 2. Details of the principle components.

	Component1	Component2	Component3
VPD	0.39497241	−0.15666692	0.905236031
Sap flow rate	0.883478921	−0.205421118	−0.421031067
Wind speed	0.251916238	0.966052608	0.057276247

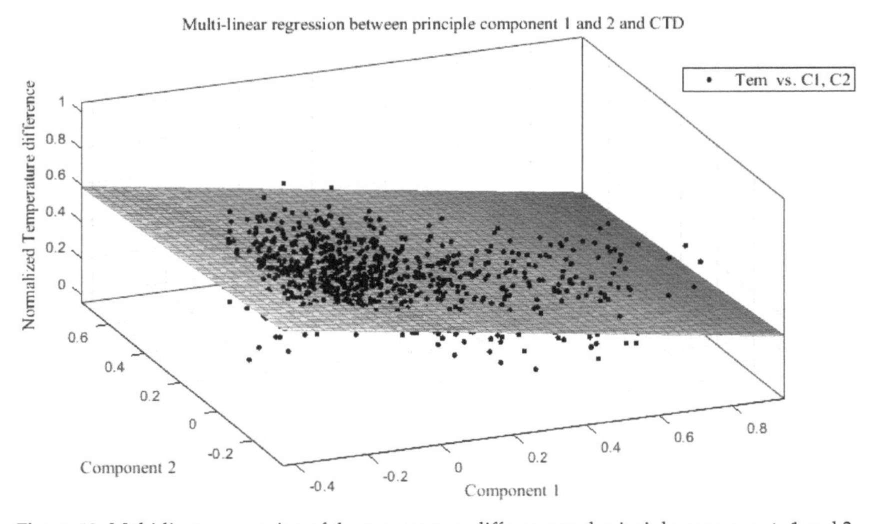

Figure 10. Multi-linear regression of the temperature difference and principle components 1 and 2.

4. Discussion

4.1. Horizontal Cooling Effect

In this experiment, the temperature difference between the outside and inside canopy was measured. The analysis indicates that the synoptic conditions in Sydney may have made a significant contribution to the results.

Changes in the transpiration rates of large trees depend largely on direct radiation changes, while regional synoptic conditions rule the wind in the studied area. The study area is close to the coastline and faces open spaces in both the eastern and the western directions, making the sea breeze the dominant factor in the surface energy balance despite the impact of downward shortwave radiation input. Further, the cooling effect of transpiration is negligible compared to the cold brought by the sea breeze. Studies conducted in Munich [13] reported that the sap flow has a significant relationship with the difference between the temperature under the canopy and that outside the canopy. Munich has a mild climate, and in the researched area, no climatic conditions (such as a stiff sea breeze) were observed. This indicates that the presence of a windy environment must be considered before using plants for transpiration cooling. The difference between the results of this study and the results in [13] indicates the high dependence of tree transpiration cooling on the environment, especially on local wind environments and under regional synoptic conditions.

In order to make a comparison between variables, the daily variation of several variables was calculated, as shown in Figure 11 (daily variation is calculated as the average at a specific hour equal to the sum of all the data measured that hour during the two weeks divided by the total number of data at that specific hour). The maximum daily temperature difference appears after around 17:00 (Figure 11d), which is earlier than the peak of the wind speed (Figure 3b) but later than the peak of the sap flow rate (Figure 11a). This indicates that the wind may exert influence on the temperature difference in the afternoon and delay the time of the peak temperature difference. At the same time, starting from 11 am, the difference between the canopy temperature and the air temperature remained at a similar level (Figure 11c). In our experiment, this also indicates that the wind is likely to dominate the energy balance and negate the effects of other factors in the local space under the canopy. If the wind is not the dominant factor, the variation in sap flow and solar radiation will lead to a significant peak on the curve in 11c after 11 am.

It should be noted that the temperature sensor outside the canopy was located in the downwind direction for most of the time in the afternoon, as shown in Figure 12. The northeast wind was significantly blocked by the large thick canopies and the layout, causing the wind in the downwind direction in the local environment to significantly weaken. This could be the reason for the results of our experiment: The wind cooled the canopy more than it cooled the outer areas in the downwind direction. Furthermore, if a sensor was placed in the upwind direction outside the canopy, the results indicate that the sensor was able to record a lower air temperature.

The PCA helped us eliminate the correlation between variables, showing that transpiration made similar contributions to cooling compared to the wind in the experiment. As mentioned above, the peaks of the daily sap flow, temperature difference, and wind appeared in sequence during the afternoon. This pattern suggests that in the early afternoon, when the daily maximum air temperature is most likely to occur, transpiration cooling plays a leading cooling role in the cooling mechanism. The transpiration cooling and cooling of the sea breeze complement each other in different periods under specific synoptic conditions, providing the city with a considerable cooling effect in the daytime and further emphasizing the critical impact of trees on the local climate.

However, it should be noted that after the principal value analysis, the correlation between the temperature difference and the principal component remained limited because the transpiration cooling of trees only accounts for a part of the cooling effect, and the shading effect is also an essential cooling mechanism of trees. Though the shading effect was not able to be analysed in the experiment, the effect of shading should not be ignored. In the daytime, changes in the incoming radiation intensity and direction can affect the temperature difference between the outside and inside of the canopy.

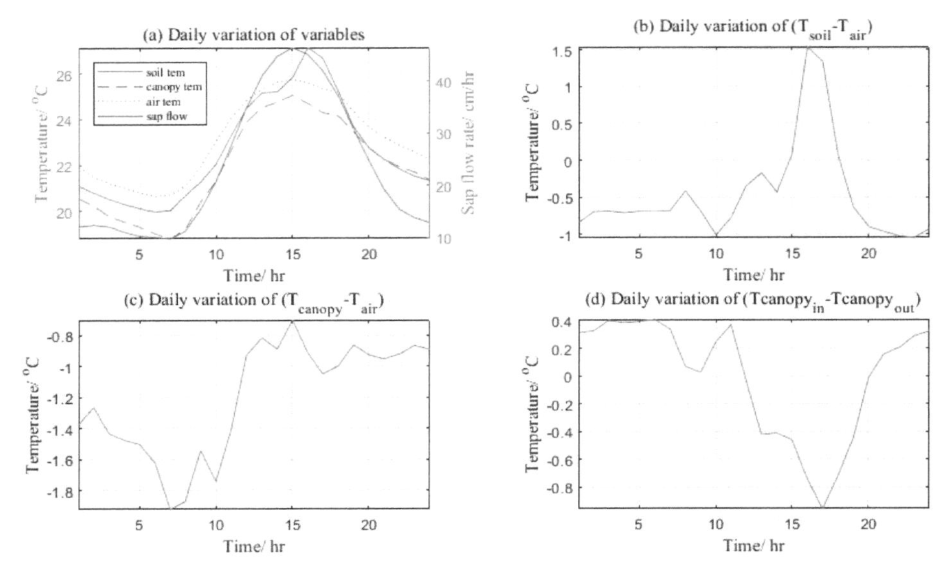

Figure 11. Daily variation of the temperature difference.

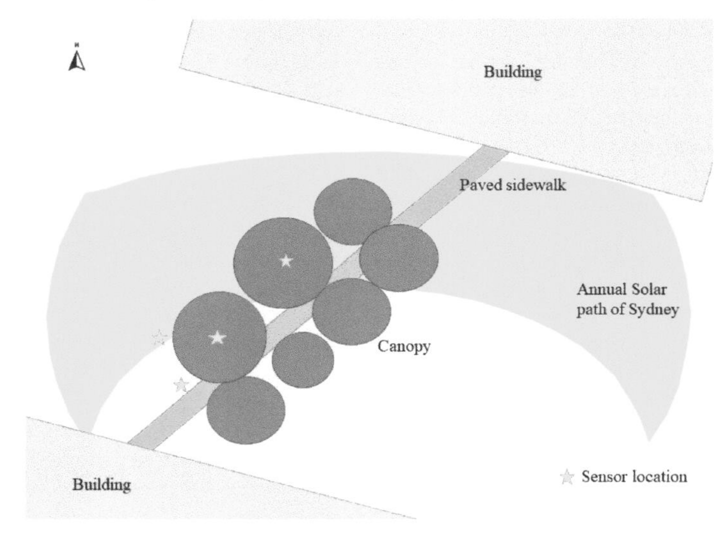

Figure 12. Canopy temperature sensor location.

Moreover, a sharp peak in the soil temperature is shown in Figure 11a. This is because the soil was receiving direct solar radiation at this moment. During the time when the the soil temperature peak occured, the recorded soil surface was not shaded by the canopy. Hence, a sharp rise in the soil surface temperature was observed. This result also underscores the importance of shading effects in tree cooling mechanisms.

4.2. Vertical Cooling Effect

In the studied area, a vertical temperature of 0.5 °C from 0.5 m to 3.5 m was observed under the canopy. This value is relatively small compared to the values in urban tree cooling studies, as mentioned above [22]. The surrounding surface conditions made the largest contributions to this result.

Different ground surfaces have different thermal properties, optical properties, and evapotranspiration capabilities. Therefore, under the same environmental conditions, different ground surfaces will have different surface temperatures and lead to a diverse vertical temperature and humidity distribution near the ground. Studies that distinguish vertical temperature differences are usually conducted on built surfaces like paved streets [22]. The warming effect of built surfaces enlarges the vertical temperature difference. This warming effect is considerable, especially when the trees are sparsely distributed. However, in our experiment, the trees are located on a dense grassland, where the soil surface remains moist most of the time. The transpiration of grass and direct evaporation from the soil cause a considerable cooling effect, and the cooling effect of the lawn and canopy work together, resulting in a more uniform vertical temperature distribution.

To confirm the cooling effect of the lawn on our measurements, a regression of the surface soil temperature and air temperature was developed, and the results are presented in Figure 13. According to the fitting equation, the projected soil temperature is significantly lower than the air temperature. When the air temperature is 30 °C, the projected soil surface temperature is 28.9 °C at the experimental site, and the temperature difference is 1.1 °C. When the air temperature is 35 °C, the projected soil surface temperature is 33.1 °C, and the temperature difference reaches 1.9 °C. This means that the cooling effect of the lawn did contribute to the vertical temperature distribution.

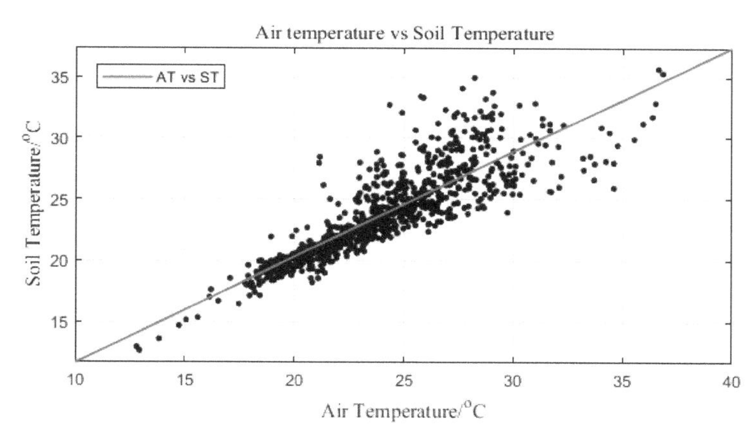

Figure 13. Correlation between the air temperature and soil temperature.

4.3. Limitation of the Research

Due to the complexity of the urban environment and the experimental conditions, the following limitations exist in this study. In this experiment, the sap flow rate was recorded but not calibrated. Hence, only the sap flow pattern was analyzed. The transpiration cooling ability is thus unable to be quantified since the transpiration cooling power cannot be calculated. Part of the data used in this experiment was taken from the nearest observation station. The distance between the experimental site and the nearest station was around 1.1 km, and the conditions of the station were similar to those of the experiment site. Since distance remained between the observation stations and the experimental site, the correlation coefficient between the station observation and site data inevitably decreased, although some studies have shown that it is feasible to use data from nearby observation stations in this kind of research. Considering the analysis results above, the influence of wind may cause a significant difference in the air temperature from the east and west sides of the canopies. However, this difference was not quantified in the present measurements since only one side of the canopy was measured. Also, the absence of radiation data can also have an impact on the multi-linear fiting results of the principle components. The R square value of the current result may be lower than the fitting result using principle components that consider radiation.

Due to the complexity of the urban environment, the results of this experiment may not be suitable for all places that have a strong coastal breeze because the local wind environment is also greatly affected by the terrain and building layout. If there are large buildings present in the upwind direction, the contribution of the wind field to the cooling of the local environment could be significantly weakened, so the role of the transpiration cooling of the plants in such scenarios is more critical. However, the shading effect produced by the possible appearance of buildings may also obfuscate the shading effect of plants. These scenarios also need to be investigated. The results of this experiment are only suitable for occasions where the local wind environment is profoundly affected by weather conditions.

5. Conclusions

The results of this study emphasize the importance of climatic conditions for the transpiration cooling effects of trees in urban overheating. We found that transpiration clearly contributes to decreasing the temperature, even under strong coastal wind conditions. Therefore, in cities, where the climate is dominated by the sea breeze, the side of the city that is far from the coastal line is less affected by the sea breeze, but transpiration still plays a critical role in the cooling of these areas. At the same time, the experiment also showed that the patterns of sea breeze and sap often do not coincide and that trees and the wind can complement each other in terms of urban overheating mitigation at different times of day. Hence, even in the areas near the coast, transpiration cooling could still be useful.

Furthermore, for cities in the desert, the situation may be opposite to the situation in this article. Dry and hot desert wind can bring heat to the city. However, the desert wind will also increase the VPD of the local space, thus amplifying the transpiration rate. Therefore, in such cities, the mitigation potential of trees planted in well-ventilated areas can be maximized. The canopy barrier can reduce the invasion of the desert wind into the city and further enhance the value of the trees in alleviating urban overheating. At the same time, it should be noted that sufficient irrigation of trees in these areas may be a precondition for effective mitigation.

At the same time, the measurement results and analysis of the vertical temperature distribution emphasize the influence of the microenvironment on the cooling effect of the trees. In a highly built urban environment, the warming effect of surfaces, such as concrete sidewalks or asphalt roads, is considerable at noon in the summer, and the warm air flow near the ground surface can be substantial. If the canopies of the trees are not large and dense enough, the warming effects of the surroundings can easily conceal the shading and transpiration cooling effects of the trees, leading to a decreased cooling effect. Therefore, in urban landscape design, to achieve effective vegetation cooling in a highly built environment, trees with a high LAI and a large canopy projection area must be selected.

Author Contributions: Conceptualization, K.G. and M.S.; Data curation, K.G. and J.F.; Formal analysis, K.G., M.S. and J.F.; Investigation, K.G.; Methodology, K.G. and M.S.; Writing—review & editing, J.F. All authors have read and agreed to the published version of the manuscript.

Funding: This research received no external funding.

Acknowledgments: This research was supported in part by the City of Parramatta with the research contract "Parramatta Urban Overheating. Thanks also for the assistance from Riccardo Paolini in the experiment preparation.

Conflicts of Interest: The authors declare no conflict of interest.

References

1. Santamouris, M.; Cartalis, C.; Synnefa, A.; Kolokotsa, D. On the impact of urban heat island and global warming on the power demand and electricity consumption of buildings—A review. *Energy Build.* **2015**, *98*, 119–124. [CrossRef]
2. Santamouris, M. Analyzing the heat island magnitude and characteristics in one hundred Asian and Australian cities and regions. *Sci. Total Environ.* **2015**, *512–513*, 582–598. [CrossRef] [PubMed]
3. Santamouris, M. Recent progress on urban overheating and heat island research. Integrated assessment of the energy, environmental, vulnerability and health impact. Synergies with the global climate change. *Energy Build.* **2020**, *207*, 109482. [CrossRef]
4. Santamouris, M.; Paolini, R.; Haddad, S.; Synnefa, A.; Garshasbi, S.; Hatvani-Kovacs, G.; Gobakis, K.; Yenneti, K.; Vasilakopoulou, K.; Gao, K.; et al. Heat Mitigation Technologies Can Improve Sustainability in Cities an Holistic Experimental and Numerical Impact Assessment of Urban Overheating and Related Heat Mitigation Strategies on Energy Consumption, Indoor Comfort, Vulnerability and Heat-Related Mortality and Morbidity in Cities. *Energy Build.* **2020**, *217*, 110002.
5. Santamouris, M.; Synnefa, A.; Karlessi, T. Using advanced cool materials in the urban built environment to mitigate heat islands and improve thermal comfort conditions. *Sol. Energy* **2011**, *85*, 3085–3102. [CrossRef]
6. Feng, J.; Gao, K.; Santamouris, M.; Shah, K.W.; Ranzi, G. Dynamic impact of climate on the performance of daytime radiative cooling materials. *Sol. Energy Mater. Sol. Cells* **2020**, *208*, 110426. [CrossRef]
7. Santamouris, M. Using cool pavements as a mitigation strategy to fight urban heat island—A review of the actual developments. *Renew. Sustain. Energy Rev.* **2013**, *26*, 224–240. [CrossRef]
8. Santamouris, M. Cooling the cities—A review of reflective and green roof mitigation technologies to fight heat island and improve comfort in urban environments. *Sol. Energy* **2014**, *103*, 682–703. [CrossRef]
9. Armson, D.; Stringer, P.; Ennos, A. The effect of tree shade and grass on surface and globe temperatures in an urban area. *Urban For. Urban Green.* **2012**, *11*, 245–255. [CrossRef]
10. Edmondson, J.L.; Stott, I.; Davies, Z.G.; Gaston, K.J.; Leake, J.R. Soil surface temperatures reveal moderation of the urban heat island effect by trees and shrubs. *Sci. Rep.* **2016**, *6*, 1–8. [CrossRef]
11. Zölch, T.; Maderspacher, J.; Wamsler, C.; Pauleit, S. Using green infrastructure for urban climate-proofing: An evaluation of heat mitigation measures at the micro-scale. *Urban For. Urban Green.* **2016**, *20*, 305–316. [CrossRef]
12. Coutts, A.M.; White, E.C.; Tapper, N.J.; Beringer, J.; Livesley, S.J. Temperature and human thermal comfort effects of street trees across three contrasting street canyon environments. *Theor. Appl. Climatol.* **2016**, *124*, 55–68. [CrossRef]
13. Rahman, M.A.; Moser, A.; Rötzer, T.; Pauleit, S. Within canopy temperature differences and cooling ability of Tilia cordata trees grown in urban conditions. *Build. Environ.* **2017**, *114*, 118–128. [CrossRef]
14. De Abreu-Harbich, L.V.; Labaki, L.C.; Matzarakis, A. Effect of tree planting design and tree species on human thermal comfort in the tropics. *Landsc. Urban Plan.* **2015**, *138*, 99–109. [CrossRef]
15. Lobell, D.B.; Bonfils, C.J.; Kueppers, L.M.; Snyder, M.A. Irrigation cooling effect on temperature and heat index extremes. *Geophys. Res. Lett.* **2008**, *35*. [CrossRef]
16. Biggs, T.W.; Scott, C.A.; Gaur, A.; Venot, J.-P.; Chase, T.; Lee, E. Impacts of irrigation and anthropogenic aerosols on the water balance, heat fluxes, and surface temperature in a river basin. *Water Resour. Res.* **2008**, *44*. [CrossRef]
17. Broadbent, A.M.; Coutts, A.M.; Tapper, N.J.; Demuzere, M. The cooling effect of irrigation on urban microclimate during heatwave conditions. *Urban Clim.* **2018**, *23*, 309–329. [CrossRef]
18. Konarska, J.; Uddling, J.; Holmer, B.; Lutz, M.; Lindberg, F.; Pleijel, H.; Thorsson, S. Transpiration of urban trees and its cooling effect in a high latitude city. *Int. J. Biometeorol.* **2016**, *60*, 159–172. [CrossRef]
19. Armson, D.; Rahman, M.A.; Ennos, A.R. A comparison of the shading effectiveness of five different street tree species in Manchester, UK. *Arboric. Urban For.* **2013**, *39*, 157–164.
20. Rahman, M.; Armson, D.; Ennos, A. A comparison of the growth and cooling effectiveness of five commonly planted urban tree species. *Urban Ecosyst.* **2015**, *18*, 371–389. [CrossRef]
21. Wang, C.; Wang, Z.-H.; Wang, C.; Myint, S.W. Environmental cooling provided by urban trees under extreme heat and cold waves in U.S. cities. *Remote Sens. Environ.* **2019**, *227*, 28–43. [CrossRef]

22. Rahman, M.A.; Moser, A.; Gold, A.; Rotzer, T.; Pauleit, S. Vertical air temperature gradients under the shade of two contrasting urban tree species during different types of summer days. *Sci. Total Environ.* **2018**, *633*, 100–111. [CrossRef] [PubMed]
23. Zweifel, R.; Böhm, J.; Häsler, R. Midday stomatal closure in Norway spruce—Reactions in the upper and lower crown. *Tree Physiol.* **2002**, *22*, 1125–1136. [CrossRef]
24. Wang, Y.; Bakker, F.; de Groot, R.; Wortche, H.; Leemans, R. Effects of urban trees on local outdoor microclimate: Synthesizing field measurements by numerical modelling. *Urban Ecosyst.* **2015**, *18*, 1305–1331. [CrossRef]
25. Tayyebi, A.; Jenerette, G.D. Increases in the climate change adaption effectiveness and availability of vegetation across a coastal to desert climate gradient in metropolitan Los Angeles, CA, USA. *Sci. Total Environ.* **2016**, *548*, 60–71. [CrossRef]
26. Pfautsch, S.; Bleby, T.M.; Rennenberg, H.; Adams, M.A. Sap flow measurements reveal influence of temperature and stand structure on water use of Eucalyptus regnans forests. *For. Ecol. Manag.* **2010**, *259*, 1190–1199. [CrossRef]
27. Dimoudi, A.; Nikolopoulou, M. Vegetation in the urban environment: Microclimatic analysis and benefits. *Energy Build.* **2003**, *35*, 69–76. [CrossRef]
28. Holden, Z.A.; Klene, A.E.; Keefe, R.F.; Moisen, G.G. Design and evaluation of an inexpensive radiation shield for monitoring surface air temperatures. *Agric. For. Meteorol.* **2013**, *180*, 281–286. [CrossRef]
29. Marshall, D. Measurement of sap flow in conifers by heat transport. *Plant Physiol.* **1958**, *33*, 385. [CrossRef]
30. Cohen, Y.; Fuchs, M.; Green, G. Improvement of the heat pulse method for determining sap flow in trees. *Plant Cell Environ.* **1981**, *4*, 391–397. [CrossRef]
31. Kendall, M.G. *The Advanced Theory of Statistics*, 2nd ed.; Charles Griffin and Co., Ltd.: London, UK, 1946.

Article

Probability Risk of Heat- and Cold-Related Mortality to Temperature, Gender, and Age Using GAM Regression Analysis

Andri Pyrgou and Mattheos Santamouris *

Faculty of Built Environment, University of New South Wales, Sydney 2052, Australia
* Correspondence: m.santamouris@unsw.edu.au

Received: 7 February 2020; Accepted: 11 March 2020; Published: 11 March 2020

Abstract: We have examined the heat and cold-related mortality risk subject to cold and heat extremes by using a generalized additive model (GAM) regression technique to quantify the effect of the stimulus of mortality in the presence of covariate data for 2007–2014 in Nicosia, Cyprus. The use of the GAM technique with multiple linear regression allowed for the continuous covariates of temperature and diurnal temperature range (DTR) to be modeled as smooth functions and the lag period was considered to relate mortality to lagged values of temperature. Our findings indicate that the previous three days' temperatures were strongly predictive of mortality. The mortality risk decreased as the minimum temperature (T_{min}) increased from the coldest days to a certain threshold temperature about 20–21°C (different for each age group and gender), above which the mortality risk increased as T_{min} increased. The investigated fixed factors analysis showed an insignificant association of gender-mortality, whereas the age-mortality association showed that the population over 80 was more vulnerable to temperature variations. It was recommended that the minimum mortality temperature is calculated using the minimum daily temperatures because it has a stronger correlation to the probability for risk of mortality. It is still undetermined as to what degree a change in existing climatic conditions will increase the environmental stress to humans as the population is acclimatized to different climates with different threshold temperatures and minimum mortality temperatures.

Keywords: heatwave; diurnal temperature range; time-series; relative risk; health

1. Introduction

The relationship between hot and cold temperatures and mortality from respiratory and cardiovascular causes is well established. Governments and scientists are concerned with the increased frequency of temperature extremes as they are associated with increased morbidity and mortality [1,2]. Exploration of time series data in different countries has revealed a different temperature threshold of their population [3], necessitating different adaptation measures for the avoidance of the climate-change impact and to increase the countries' capacity to function at a forthcoming temperature [4–8]. It is of utmost importance for the threshold temperature to be determined per country or prevalent climate as there is a gap in the identification of the correct course of adaptation.

Mortality risk with respect to temperature has been assessed by scientists in northeastern Europe [9–12], the USA [13,14], and China [4] by considering fixed variables such as gender and age, resulting in contrasting findings of the threshold temperature and minimum mortality temperature (MMT). MMT is defined as the temperature at which there is the lowest risk of mortality according to a probability risk assessment. The question of whether fixed factors are affecting the MMT has to be addressed carefully in the eastern Mediterranean region because of the limited number of studies showing high MMT; 29–32 °C [15]. Generally, MMT has been found to be lower for populations living in colder climates and higher for populations living in warmer climates [11,14,16] with the temperature–mortality relationship described as a J-, V-, or U-shaped curve.

People acclimatize to new temperatures at varying rates and to a certain extent based on physiological parameters such as age, gender, and other prevalent health conditions. The environmental stress even within a day may be a factor of increased cardiovascular and respiratory mortality. Moreover, the effect of exposure to extreme cold or heat conditions is not limited to the specific day, but may be delayed in time [2,14,17]. The diurnal temperature range (DTR) is defined as the difference between the daily minimum and maximum temperatures, with some studies correlating a high DTR with an increase of mortality risk [17].

To fill the above research gaps, we addressed the issue of temperature-related mortality via the use of the generalized additive model (GAM) regression technique. Using the GAM, we investigated the effect of same day temperatures and the weighted average temperature of the preceding three on mortality rates. The use of the GAM model to examine the short-term mortality relationship to regional minimum temperatures, maximum temperatures, and DTR variations revealed that MMT should be calculated using daily minimum temperature values.

2. Methods

2.1. Study Area and Datasets

Hourly weather data (temperature [°C]) and daily mortality data for 2007 to 2014 inclusive were collected for two meteorological stations [18] in the urban (35.17°N, 33.36°E) and rural (35.05°N, 33.54°E) areas of Nicosia, Cyprus. Nicosia is the capital of the island of Cyprus, located in the eastern basin of the Mediterranean Sea with a hot summer Mediterranean climate and hot semi-arid climate (in the northeastern part of island), according to the Köppen climate classification signs Csa (Mediterranean hot summer climates) and BSh (Hot semi-arid climates) [19], with warm to hot dry summers and wet winters.

The daily mortality counts were gathered only for circulatory and respiratory causes of death and included ischemic heart diseases (I20–I25), cerebrovascular diseases (I60–I69), other heart diseases (I30–I51), other circulatory diseases (I00–I15, I26–I28, I70–I99), influenza (J00–J99), pneumonia (J12–J18), chronic lower respiratory diseases (J40–J47), and other respiratory causes (J00–J06, J20–J39, J60–J99), according to the ICD-10-CM (International Classification of Diseases, Tenth Revision, Clinical Modification). The daily mortality data were provided by the Health Monitoring Unit of the Ministry of Health of Cyprus.

2.2. Log-Linear Regression of Mortality–Temperature Relation

The log mortality based on temperature was assumed to be smooth, but not necessarily linear and a generalized additive model (GAM) was used. This GAM offered a high quality of prediction of the dependent variable (log mortality rate) from the various distributions by estimating unspecific (non-parametric) functions of the predictor variables xj, which were connected to the dependent variable (mortality rate) via a link function.

As the death on a given day is not only a function of the same-day exposure to temperature but is also affected by exposure during a certain lag period, we also used the weighted average temperature of the preceding three days prior to the death. The lag period was determined using the cross correlation function (CCF) in RStudio software. Cross-correlation analysis showed the similarity of two series as a function of the displacement of one relative to the other. Figure 1 shows the cross correlation of mortality rate with minimum daily temperature for a cold period (months NDJFMA) and a hot period (months MJJASO). Similar analysis was also done for the mean and the maximum daily temperatures. The mortality rate decreased with increasing temperature during the cold period (Figure 1a) with a lag period of four days (that is, the highest peak of mortality four days after the coldest temperature), whereas the mortality rate reached a peak on the same day (no lag period) as the maximum temperature (Figure 1b). The lag period of cross correlation function (CCF analysis) that had an effect on mortality rate was four days during the cold period and 0 days during the hot period. In an individual, the lag period should not be regarded as a well-defined interval as it may vary according to the magnitude of the temperature and individual characteristics such as acclimatization habits to heat, genetic background, and income.

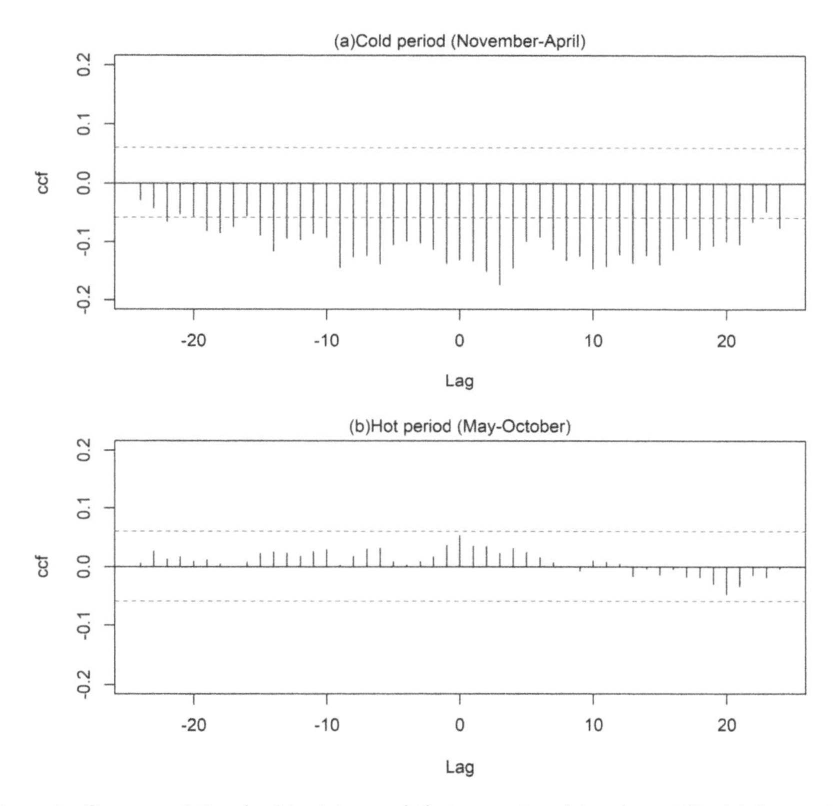

Figure 1. Cross correlation for (**a**) minimum daily temperature (x) and mortality (y) for months November to April; (**b**) Minimum daily temperature (x) and mortality (y) for months May to October). Dotted blue horizontal lines show the 95% significance limits.

To address the lagged dependence of mortality on temperature, the same-day maximum (T_{max}) and minimum (T_{min}) temperatures, the same-day diurnal temperature range (DTR), and the average temperature of the preceding three days ($T_{max}[-3]$ and $T_{min}[-3]$, respectively) and weighted average DTR of the preceding three days (DTR[−3]) were used in order to closer examine the lag period of four days found during the cold period. Diurnal temperature range is the temperature difference of the daily maximum value and the daily minimum value [20].

The advantage of GAM was to limit the error in the prediction of the dependent variable—mortality rate—by defining the model in terms of smooth function. In the GAM, the degree of smoothness of the estimated mortality–temperature relative risk curve is controlled by its number of degrees of freedom (df). Many degrees of freedom were preferred to allow highly nonlinear shapes.

$$\text{Relative risk of Mortality} = \beta + f_1(x_1) + f_2(x_2) + \ldots + f_m(x_m) \quad (1)$$

The functions fj may have a parametric or non-parametric form. xj represents the temperature term, which could be T_{min}, T_{max}, $T_{max}[-3]$, $T_{min}[-3]$, DTR or DTR[−3], as explained above. The relative risk of mortality was calculated using the gam function of the mgcv package in RStudio software [21]. The limitation of this short-term temperature analysis was that longer-term population characteristics were not considered such as health behaviors (smoking, drinking), comorbidities (hypertension, diabetes, cancer, etc.), medications, and access to health care.

3. Results

Temperature–Mortality Relative Risk Analysis

Figure 2 shows the temperature–mortality relative risk function estimated for Nicosia using GAM analysis. We examined six temperature parameters: the maximum daily temperature (T_{max}), the weighted average of the maximum daily temperatures of the preceding three days ($T_{max}[-3]$), the minimum daily temperature (T_{min}), the weighted average of the minimum daily temperatures of the preceding three days ($T_{min}[-3]$), the diurnal temperature range (DTR) of the day, and the weighted average of the DTR of the preceding three days (DTR[−3]). A smooth function of time with 50 df over the investigated years was used for the model. Similar findings were also found for smaller and larger df.

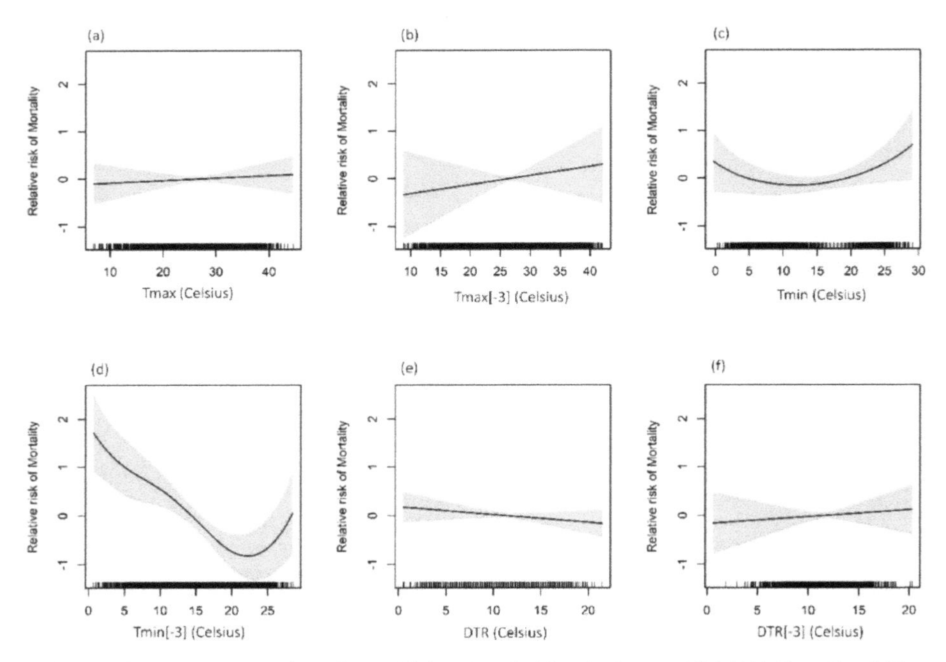

Figure 2. Temperature–mortality relative risk functions for Nicosia, Cyprus, 2004–2014. T_{max}, $T_{max}[-3]$, T_{min}, $T_{min}[-3]$, DTR, and DTR[−3] are shown in (**a**–**f**), respectively.

The relationship between maximum daily temperature and probability risk of mortality seems linear, with a steeper slope when the average of the preceding three days' maximum temperature is considered. The analysis of the minimum daily temperature with respect to the mortality's probability risk (Figure 2c,d) showed a U shape for same-day relationship and an inverse J shape when $T_{min}[-3]$ was used. That is, mortality risk decreased as the minimum temperature increased from the coldest temperatures and began to rise as the temperature increased from a certain threshold temperature (approximately 22 °C). The applied model offers flexibility and agrees with Curriero et al. [14] as the population in a warm region (such as Nicosia) tends to be more vulnerable to cold rather than those residing in cold climates who are most sensitive to heat.

The analysis of DTR (Figure 2d,f) showed no significant results (mortality relative risk close to zero), insinuating that the human body could adapt to any DTR within the same day. These results thus need to be interpreted with attention, as previous studies have reported noteworthy findings of DTR with the probability risk of mortality, therefore a more focused analysis sub-grouped by age could elaborate on better assumptions.

Further analysis focused on the T_{min} of the current day and the weighted average of the preceding three days ($T_{min}[-3]$) by subgrouping the risk by gender and age groups. The variables gender and age were related using GAM with these minimum temperatures and the results are shown in Figure 3a,b. According to Figure 3a,b, cold temperatures impose a greater risk than hot temperatures, but other factors such as respiratory epidemics, usually present in winter, made unclear the exact role of temperatures on increased mortality.

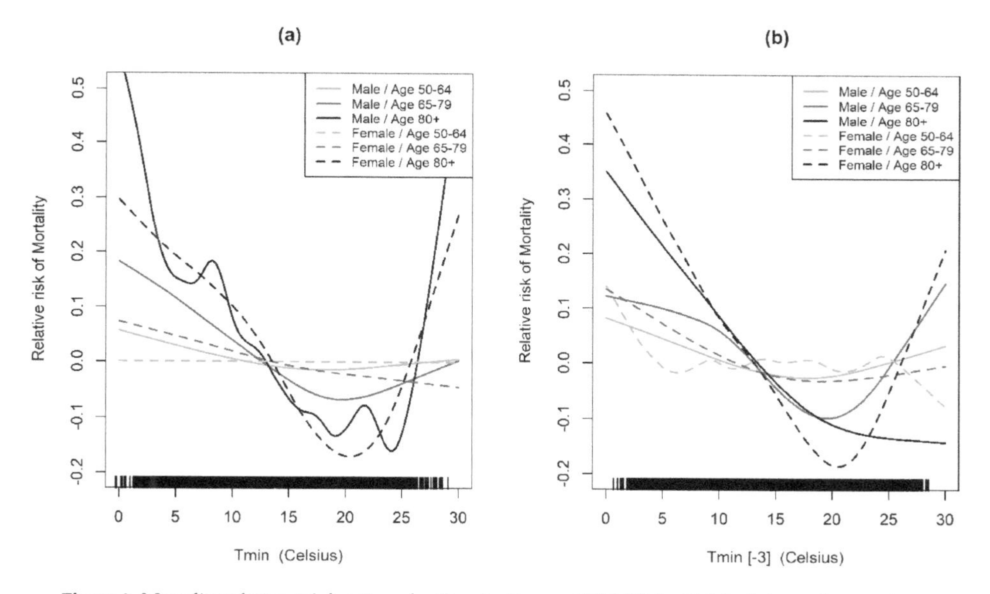

Figure 3. Mortality relative risk functions for Nicosia, Cyprus, 2007–2014 with (**a**) minimum temperature of the same day (T_{min}), (**b**) Weighted average minimum temperature of the preceding three days ($T_{min}[-3]$).

On the other hand, high minimum temperatures and heatwaves were also associated with increased mortality. Heat waves have gained more attention due to the urban warming attributed to greenhouse gases and other anthropogenic sources. Other studies have shown that different cities have different sensitivities to extremes in temperature and that the latitude and local climate are factors to consider [14,22]. Air conditioning and human behavior can substantially modify the adverse effects of high temperatures, but even in the hot city of Nicosia, where people are more accustomed to higher temperatures and use air conditioning frequently, the effect of heat on health showed increasing deaths during heat waves. Thus, adaptation should be readdressed and governments should aim for adequate people awareness.

Figure 4 shows the relative risk of mortality for the DTR of the same day and the weighted average DTR of the preceding three days. Kan et al. (2007) hypothesized that large diurnal temperature change might be a source of additional environmental stress, leading to a greater risk factor for death [17].

In contrast, our results showed a greater risk at DTR in the range of 6–8 °C, and smaller risk at larger DTRs. Men and women aged 50–64 (magenta lines) were not affected by the variations in DTR throughout the day or the preceding three days. Men 65 years and over had a greater relative risk for DTR smaller than 5 °C, whereas for larger DTR, there was a negative relative risk, showing that large DTR did not impose a risk factor for death in Nicosia. Overall, we found that DTR was independently associated with daily mortality in Nicosia and that fluctuations in DTR appeared to mostly affect people over 80, probably because they have reduced ability to regulate body temperatures, thus making them marginally more vulnerable.

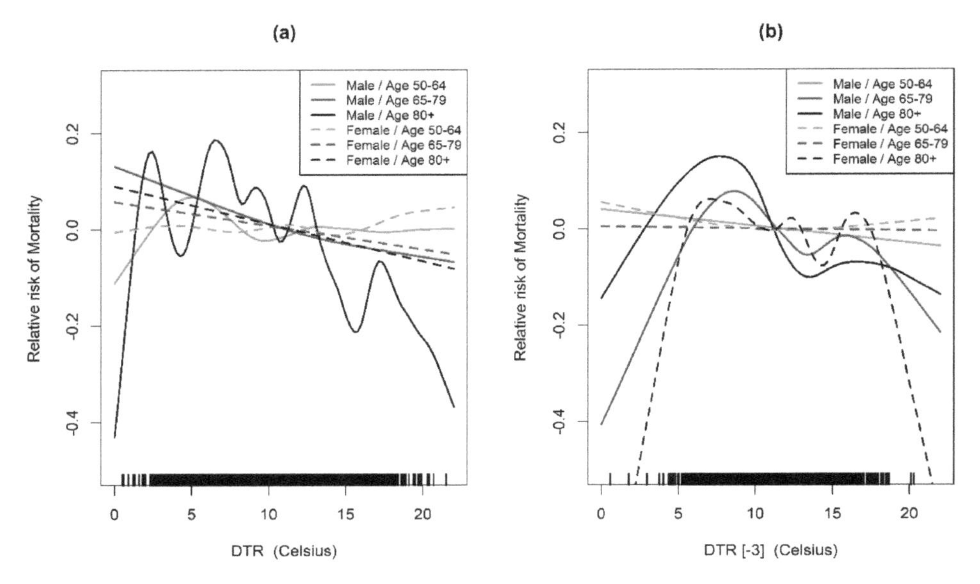

Figure 4. Mortality relative risk functions for Nicosia, Cyprus, 2007-2014 with (**a**) DTR of the same day, (**b**) Weighted average DTR of the preceding three days (DTR[−3]).

4. Discussion

The mortality risk by respiratory and circulatory causes for different age groups and per gender in relation to temperature has been examined. We employed two regression techniques to evaluate the impact of fixed and time-dependent factors on the vulnerability to temperature–mortality associations for 6882 deaths resulting from cardiovascular and respiratory causes between the years 2007 and 2014, in Nicosia, Cyprus. To model the relationship between temperature and mortality, we used the generalized additive model (GAM) to assess the interaction between a variable and the observation time and to interpret quantitative results. The GAM model revealed significantly increased mortality on hot (no lag period) and cold days (with a lag of 3–4 days), agreeing with a similar study in Estonia [4,9]. The adverse effects of heat on health are usually more direct with increased mortality on the same day or a couple of days after a heatwave [2,10]. As proposed by Curriero et al. [14], adaptation of populations to their local climate is evident by the increased health risk in relation to cold temperatures in warmer climates and on the contrary in relation to high temperature in colder climates. In warmer climates such as Nicosia, the people are more acclimatized to high temperature conditions and therefore, a steeper increase of the relative risk of mortality in colder conditions was observed.

The most important result to emerge from the analysis is that in the investigated area, the threshold temperature was about 21 °C, supporting the assumption that people in the area are more acclimatized to higher temperatures. MMT has been calculated in over 400 locations using the mean daily temperatures showing values of 29–32 °C in the eastern Mediterranean region and over 23 °C in the rest of the Mediterranean [15]. The probability risk of mortality using GAM revealed a stronger relation with the minimum daily temperature, so future studies should focus on the investigation of MMT using daily minimum temperatures. A similar GAM analysis in Shanghai [17] found smaller relative risk for elderly people with values of 0.3, therefore in Nicosia, where the population is more accustomed to higher temperatures, they are more vulnerable to lower temperatures with relative risk up to 0.5.

Our cross-correlation results agreed with previous studies where the effects of heat on mortality rate shortly after temperatures start to increase, whereas the effects of cold may take longer to emerge, and, depending on the latitude and the local climate, these periods may vary [2,9,23]. The results are in complete agreement with previous studies [24], finding no difference in heat-mortality risk between men and women, and weak evidence of a higher association of cold-mortality risk for men. The results are also in line with a study in Nicosia and a study in Stockholm [11], which found a difference in heat wave duration effects by age groups [2], while the increasing susceptibility to cold temperatures in the elderly has not been shown before.

This study has not confirmed previous research on DTR. In fact, in contrast with what was previously thought, we found that DTR does not have a noteworthy effect on mortality risk. This serves to allow for more focused research on environmental stress factors and whether prolonged duration of extreme heat or cold conditions is more important than within day variations of temperature.

A limitation of this study is the poor correlation of indoor and outdoor temperatures due to a number of modifying factors such as air conditioning, ventilation, and clothing. Time indoors affects the individual's exposure as well as workplace conditions and other comorbidities. Another limitation is that we did not adjust the analysis for micro-level socio-economic or demographic variables or other comorbidities that could have a potential confounding or modifying effect on the mortality–temperature relationship.

5. Conclusions

The main concern of the paper was to examine the mortality risk in relation to high or low temperatures of different age groups and compare them between the two genders. Mortality risk has only been evaluated for respiratory and circulatory causes. Particular attention is paid to elderly people, over 65 years old, as the results have shown a great vulnerability to ambient air temperature. We have addressed not only minimum and maximum daily temperatures, but also the diurnal temperature range (DTR) in order to examine the sensitivity for within the same day air temperature variations.

The originality of our approach lies in the fact that we have combined the cross correlation analysis to identify the effect of the preceding days' temperature with the generalized additive model (GAM) regression technique. From the research that has been performed, it is possible to conclude that there was increased mortality on extremely hot and cold days. The effects of the heat had no lag period, whereas cold effects had a lag effect of 3–4 days. The existence of these responses implies that in warmer climates, people are more acclimatized to high temperatures, and therefore a higher mortality risk was observed at colder temperatures with a lag of three days.

The approach used in this paper is applicable to several environmental areas such as air quality analysis where the results may be subjective to a delay period and may slowly diminish human health and well-being. The results of this study should alert organizations and governments on the possible impacts of climate change on public health by not considering adaptation. The identification of a threshold temperature per latitude and local climate will assist in the evaluation of the adaptation capacity of a specific population. This threshold temperature, according to our results, should be calculated using the daily minimum temperature. Nevertheless, even if humans become fully acclimatized to high temperatures, their health may still be negatively affected as a result of the poorer air quality associated with extremely high temperatures [2,25].

On the basis of the promising findings presented in this paper, work on the remaining issues is continuing to examine whether socio-economic or demographic variables or other comorbidities could have a potential confounding or modifying effect on the mortality–temperature relationship.

Author Contributions: M.S. conceived the research topic. A.P. obtained the datasets, created the figures, and analyzed the results. Both authors (A.P. and M.S.) contributed in the discussion of the results and reviewed the manuscript. All authors have read and agreed to the published version of the manuscript.

Acknowledgments: The authors are grateful to the Ministry of Agriculture, Rural Development, and Environment (MADRE) of the Republic of Cyprus for the Department of Meteorology historical meteorological data, and the Health Monitoring Unit of the Ministry of Health of Cyprus for the mortality data. The ideas and opinions expressed herein are those of the authors. Endorsement of these ideas and opinions by the Ministry of Health of Cyprus is not intended nor should it be inferred.

Conflicts of Interest: The authors declare no conflicts of interest.

References

1. Hondula, D.M.; Balling, R.C.; Vanos, J.K.; Georgescu, M. Rising Temperatures, Human Health, and the Role of Adaptation. *Curr. Clim. Chang. Rep.* **2015**, *1*, 144–154. [CrossRef]
2. Pyrgou, A.; Santamouris, M. Increasing Probability of Heat-Related Mortality in a Mediterranean City Due to Urban Warming. *Int. J. Environ. Res. Public Health* **2018**, *15*, 1571. [CrossRef] [PubMed]
3. Coffel, E.D.; Horton, R.M.; De Sherbinin, A. Temperature and humidity based projections of a rapid rise in global heat stress exposure during the 21st century. *Environ. Res. Lett.* **2018**, *13*. [CrossRef]
4. Chen, R.; Yin, P.; Wang, L.; Liu, C.; Niu, Y.; Wang, W.; Jiang, Y.; Liu, Y.; Liu, J.; Qi, J.; et al. Association between ambient temperature and mortality risk and burden: Time series study in 272 main Chinese cities. *BMJ* **2018**, *363*. [CrossRef] [PubMed]
5. Rocklöv, J.; Forsberg, B.; Ebi, K.; Bellander, T. Susceptibility to mortality related to temperature and heat and cold wave duration in the population of Stockholm County, Sweden. *Glob. Health Action* **2014**, *7*, 1–11. [CrossRef] [PubMed]
6. Gronlund, C.J.; Berrocal, V.J.; White-Newsome, J.L.; Conlon, K.C.; O'Neill, M.S. Vulnerability to extreme heat by socio-demographic characteristics and area green space among the elderly in Michigan, 1990–2007. *Environ. Res.* **2015**. [CrossRef]
7. Goggins, W.B.; Ren, C.; Ng, E.; Yang, C.; Chan, E.Y.Y. Effect modification of the association between meteorological variables and mortality by urban climatic conditions in the tropical city of Kaohsiung. Taiwan. *Geospatial Health* **2013**. [CrossRef]
8. Milojevic, A.; Armstrong, B.G.; Gasparrini, A.; Bohnenstengel, S.I.; Barratt, B.; Wilkinson, P. Methods to estimate acclimatization to urban heat island effects on heat-and cold-related mortality. *Environ. Health Perspect.* **2016**, *124*, 1016–1022. [CrossRef]
9. Orru, H.; Åström, D.O. Increases in external cause mortality due to high and low temperatures: Evidence from northeastern Europe. *Int. J. Biometeorol.* **2017**, *61*, 963–966. [CrossRef]
10. Åström, D.O.; Åström, C.; Rekker, K.; Indermitte, E.; Orru, H. High summer temperatures and mortality in Estonia. *PLoS ONE* **2016**, *11*, 1–10. [CrossRef]
11. Oudin Åström, D.; Tornevi, A.; Ebi, K.L.; Rocklöv, J.; Forsberg, B. Evolution of minimum mortality temperature in Stockholm, Sweden, 1901–2009. *Environ. Health Perspect.* **2016**, *124*, 740–744. [CrossRef] [PubMed]
12. Huynen, M.M.T.E.; Martens, P.; Schram, D.; Weijenberg, M.P.; Kunst, A.E. The impact of heat waves and cold spells on mortality rates in the Dutch population. *Environ. Health Perspect.* **2001**, *109*, 463–470. [CrossRef]
13. Anderson, B.G.; Bell, M.L. Weather-related mortality: How heat, cold, and heat waves affect mortality in the United States. *Epidemiology* **2009**, *20*, 205. Available online: http://www.pubmedcentral.nih.gov/articlerender.fcgi?artid=3366558&tool=pmcentrez&rendertype=abstract (accessed on 15th December 2019). [CrossRef]
14. Curriero, F.C.; Heiner, K.S.; Samet, J.M.; Zeger, S.L.; Strug, L.; Patz, J.A. Temperature and mortality in 11 cities of the eastern United States. *Am. J. Epidemiol.* **2002**. [CrossRef] [PubMed]
15. Yin, Q.; Wang, J.; Ren, Z.; Li, J.; Guo, Y. Mapping the increased minimum mortality temperatures in the context of global climate change. *Nat. Commun.* **2019**, *10*, 4640. [CrossRef] [PubMed]

16. Baccini, M.; Biggeri, A.; Accetta, G.; Kosatsky, T.; Katsouyanni, K.; Analitis, A.; Anderson, H.R.; Bisanti, L.; D'Ippoliti, D.; Danova, J.; et al. Heat effects on mortality in 15 European cities. *Epidemiology* **2008**. [CrossRef] [PubMed]
17. Kan, H.; London, S.J.; Chen, H.; Song, G.; Chen, G.; Jiang, L.; Zhao, N.; Zhang, Y.; Chen, B. Diurnal temperature range and daily mortality in Shanghai, China. *Environ. Res.* **2007**, *103*, 424–431. [CrossRef]
18. Republic of Cyprus. Department of Meteorology, Cyprus. Available online: http://www.moa.gov.cy/moa/ms/ms.nsf/DMLannual_en/DMLannual_en?OpenDocument (accessed on 25th April 2019).
19. Peel, M.C.; Finlayson, B.L.; McMahon, T.A. Updated world map of the Koppen-Geiger climate classification. *Hydrol. Earth Syst. Sci.* **2007**, *11*, 1633–1644. [CrossRef]
20. Pyrgou, A.; Santamouris, M.; Livada, I. Spatiotemporal Analysis of Diurnal Temperature Range: Effect of Urbanization, Cloud Cover, Solar Radiation, and Precipitation. *Climate* **2019**, *7*, 89. [CrossRef]
21. Wood, S. Mgcv: Mixed GAM Computation Vehicle with Automatic Smoothness Estimation. Available online: https://cran.r-project.org/web/packages/mgcv/index.html (accessed on 20 October 2019). [CrossRef]
22. Braga, A.L.F.; Zanobetti, A.; Schwartz, J. The effect of weather on respiratory and cardiovascular deaths in 12 U.S. cities. *Environ. Health Perspect.* **2002**, *110*, 859–863. [CrossRef]
23. Brooke Anderson, G.; Bell, M.L. Heat waves in the United States: Mortality risk during heat waves and effect modification by heat wave characteristics in 43 U.S. communities. *Environ. Health Perspect.* **2011**, *119*, 210–218. [CrossRef] [PubMed]
24. Son, J.-Y.; Liu, J.C.; Bell, M.L. Temperature-related mortality: a systematic review and investigation of effect modifiers. *Environ. Res. Lett.* **2019**, *14*, 073004. [CrossRef]
25. Pyrgou, A.; Hadjinicolaou, P.; Santamouris, M. Enhanced near-surface ozone under heatwave conditions in a Mediterranean island. *Sci. Rep.* **2018**, *8*, 9191. [CrossRef] [PubMed]

Article

The Effects of Historical Housing Policies on Resident Exposure to Intra-Urban Heat: A Study of 108 US Urban Areas

Jeremy S. Hoffman [1,2,*], Vivek Shandas [3] and Nicholas Pendleton [1,2]

[1] Science Museum of Virginia, Richmond, VA 23220, USA; pendletonnv@mymail.vcu.edu
[2] Center for Environmental Studies, Virginia Commonwealth University, Richmond, VA 23220, USA
[3] Nohad A. Toulan School of Urban Studies and Planning, Portland State University, Portland, OR 97201, USA; vshandas@pdx.edu
* Correspondence: jhoffman@smv.org

Received: 5 November 2019; Accepted: 3 January 2020; Published: 13 January 2020

Abstract: The increasing intensity, duration, and frequency of heat waves due to human-caused climate change puts historically underserved populations in a heightened state of precarity, as studies observe that vulnerable communities—especially those within urban areas in the United States—are disproportionately exposed to extreme heat. Lacking, however, are insights into fundamental questions about the role of historical housing policies in cauterizing current exposure to climate inequities like intra-urban heat. Here, we explore the relationship between "redlining", or the historical practice of refusing home loans or insurance to whole neighborhoods based on a racially motivated perception of safety for investment, with present-day summertime intra-urban land surface temperature anomalies. Through a spatial analysis of 108 urban areas in the United States, we ask two questions: (1) how do historically redlined neighborhoods relate to current patterns of intra-urban heat? and (2) do these patterns vary by US Census Bureau region? Our results reveal that 94% of studied areas display consistent city-scale patterns of elevated land surface temperatures in formerly redlined areas relative to their non-redlined neighbors by as much as 7 °C. Regionally, Southeast and Western cities display the greatest differences while Midwest cities display the least. Nationally, land surface temperatures in redlined areas are approximately 2.6 °C warmer than in non-redlined areas. While these trends are partly attributable to the relative preponderance of impervious land cover to tree canopy in these areas, which we also examine, other factors may also be driving these differences. This study reveals that historical housing policies may, in fact, be directly responsible for disproportionate exposure to current heat events.

Keywords: urban heat islands; environmental justice; climate change; redlining

1. Introduction

No other category of hazardous weather event in the United States has caused more fatalities over the last few decades than extreme heat [1]. In fact, extreme heat is the leading cause of summertime morbidity and has specific impacts on those communities with pre-existing health conditions (e.g., chronic obstructive pulmonary disease, asthma, cardiovascular disease, etc.), limited access to resources, and the elderly [2–4]. Excess heat limits the human body's ability to regulate its internal temperature, which can result in increased cases of heat cramps, heat exhaustion, and heatstroke and may exacerbate other nervous system, respiratory, cardiovascular, genitourinary, and diabetes-related conditions [5]. As heat extremes in urban areas become more common, longer in duration, and more intense across the US and globe [6,7] due to unmitigated human emissions of heat-trapping gases from fossil fuels [8] as well as urban expansion [9], the number of deaths and attendant illnesses are expected to increase around the US [10].

Urban landscapes amplify extreme heat due to the imbalance of low-slung built surfaces to natural, non-human manufactured landscapes [11,12]. This urban heat island effect can cause temperatures to vary as much as 10 °C within a single urban area [13], even without comparison to a "traditional" rural baseline for assessing UHI. Others, including Li et al. (2017), found that the density of total impervious surface area (ISA) is a major predictor for land surface temperatures, or the "surface urban heat island" studied here [14]; yet others describe the apparent cooling effects of urban green spaces. In general, greenspace, trees, or water bodies within a city have been correlated with cooler land surface temperatures (LST), and more greenspace or water is related to lower urban LST at the location of that greenspace [15–18]. Hamstead et al. (2016) studied the role of landscape composition on surface temperatures by dividing New York City into 22 classes at 3-m resolution and identifying the specific ranges of land surface temperature represented within each class [19]. The authors conclude that urban areas contain discernable "classes" of form—the integration of land use and land cover—and that those sets have "distinct temperature signatures".

Emerging research suggests that many of the hottest urban areas also tend to be inhabited by resource-limited residents and communities of color [20,21], underscoring the emerging lens of environmental justice as it relates to urban climate change and adaptation. In one study, Voelkel and others (2018) found that residents living in neighborhoods with higher racial diversity, extreme poverty, and lower levels of formal education were statistically more likely to be exposed to greater heat—the neighborhood heat effect [21]. Still other studies have found that those with the least access to resources, more advanced in age, and people with pre-existing face some of the greatest burden [22]. While the evidence about the distributional implications of heat waves mounts, we still do not have a clear uniting principle to explain consistent patterns between an emerging challenge like intra-urban heat and observable records of excess mortality and morbidity among underserved populations. If heat varies across urban environments, then why are communities of color and resource-limited communities living in the hottest areas? Could a plausible explanation be the presence of past urban planning programs and housing policies that have heightened disproportionate exposure to intra-urban heat in US cities?

The present study further examines the relationship between present-day spatial patterns of inequitable exposure to intra-urban heat and historical housing policies, which were applied to many US cities in the early 20th century. We specifically examine maps generated by the Home Owners' Loan Corporation's (HOLC) practice of "redlining" [23,24] in the 1930s. As part of a national program to lift the US out of a recession, HOLC refinanced mortgages at low interest rates to prevent foreclosures, and in the process created color-coded residential maps of 239 individual US cities with populations over 40,000. HOLC maps distinguished neighborhoods that were considered "best" and "hazardous" for real estate investments (largely based on racial makeup), the latter of which was outlined in red, leading to the term "redlining." These "Residential Security" maps reflect one of four categories ranging from "Best" (A, outlined in green), "Still Desirable" (B, outlined in blue), "Definitely Declining" (C, outlined in yellow), to "Hazardous" (D, outlined in red), relating directly to subsequent access to mortgage lending and at least partially to the racial makeup of that neighborhood.

Though redlining was banned in the US as part of the Fair Housing Act of 1968, a majority of those areas deemed "hazardous" (and subsequently "redlined") remain dominantly low-to-moderate income and communities of color, while those deemed "desirable" remain predominantly white with above-average incomes [24]. Those living in redlined areas experienced reduced credit access and subsequent disinvestment, leading to increased segregation and lower home ownership, value, and personal credit scores, even when compared to those similar-sized US cities that did not receive a HOLC map [23]. Increasingly evident is the legacy of these historic policies in racial disparities in health care, access to healthy food, incarceration, resources allotted for schools, and public infrastructure investment such as the privileging of the suburban highway system at the expense of the city's public transportation [25].

Similarly, as areas that received severely limited real estate investment over time, we might expect those areas to have fewer environmental amenities that help to clean and cool the air, including urban tree canopy [26]. Recent studies describe the increased likelihood that those who are poor and communities of color are more likely living in areas with fewer trees and poorer air quality [27,28]. At the same time, the extent to which these policies may have resulted in environmental disparity as a consequence of systematic disinvestment nationally largely remains an open question. We seek here to assess if evidence of disproportionate environmental stressors (specifically anomalous urban land surface temperatures) exists through the lens of these long-term housing policies, and if a national-scale signal varies by region in the US.

By assessing HOLC maps from aggregated urban areas in the United States (Figure 1) in relation to the relative anomaly of land surface temperature within and outside redlined areas, we ask two questions: (1) do historical policies of redlining help to explain current patterns of exposure to intra-urban heat in US cities? and (2) how do these patterns vary by geographic location of cities? Our intent is not to explain why precisely these patterns exist; instead, we seek to describe their relation through spatial analysis of historical redlining maps and present-day warm season intra-urban land surface temperature anomalies. By examining these patterns, we aim to assess how current patterns of intra-urban heat inequities may result from a combination of historical policies that may be further exacerbated by present-day planning practices that fail to center communities that have been historically underserved in adaptation and mitigation of these patterns.

Figure 1. Map of 108 US cities with HOLC Residential Security maps included in this study. These areas may include several smaller-area HOLC maps that have been aggregated into a larger urban area (Supplementary Materials I).

2. Materials and Methods

We use the University of Richmond's Digital Scholarship Lab's "Mapping Inequality" database (Figure 2a, Richmond, VA, USA, [29]) to download each available city's HOLC map shapefile individually (n = 239). To make analysis of Landsat-derived LST maps less computationally complex, we then condense the 239 unique HOLC maps into a database of 108 US cities or urban areas that overlap within Landsat 8 imagery tiles, and excluding any cities that were not mapped with at least one of all four HOLC security rating categories (n = 4). In some cases, HOLC map shapefile boundaries needed to remove overlapping security rating boundaries, boundary crossings over bodies of water, and to merge overlapping maps drawn in the same generalizable urban area and/or because they were drawn during different years (Supplementary Materials I).

We assess patterns of intra-urban land surface temperatures in the 108 HOLC areas using readily-accessible Landsat 8 satellite-derived northern hemisphere summertime (June–August) land surface temperatures (LSTs) following accepted United States Geological Survey calculation protocol (30 × 30 m resolution, TIRS Band 10, Normal Difference Vegetation Index [NDVI] emissivity corrected LST, Figure 2b [20,30,31]. This LST method relies on transforming raw Landsat 8 TIRS Band 10 data into top-of-atmosphere spectral radiance and then into at-sensor brightness temperatures. LST is then calculated by correcting the at-sensor brightness temperatures by surface emissivity calculated from the NDVI (derived from Bands 4 and 5 [30]). LST maps were only generated from imagery that satisfied a threshold for less than 10 percent scene cloud coverage and had to have been collected in the northern hemisphere summertime between 2014 and 2017. While these LST descriptions of intra-urban heat are coarse in spatial resolution and not the most representative of human-level, experiential air temperatures which are better resolved by dense networks of air temperature and humidity monitors [13,32], LST maps such as these have been widely applied to questions of large-scale patterns related to urban land use and heat-related public health outcomes for individual US cities [20,33].

We then use Zonal Statistics in ESRI's ArcGIS Spatial Analyst toolbox to estimate the mean of the derived Landsat 8 LSTs within each individual HOLC security rating polygon within a given urban area (e.g., Figure 2b,c). We then estimate each individual HOLC security rating polygon's land surface temperature anomaly from the area-wide mean LST from all HOLC security rating polygons (referred to as δLST, Equation (1)).

$$\delta \mathrm{LST}_{area,\ polygon} = \overline{\mathrm{LST}_{area,\ polygon}} - \overline{\mathrm{LST}_{area,\ all\ polygons}} \quad (1)$$

This δLST estimate gives us the ability to show relatively how much warmer or cooler a particular HOLC security rating polygon is from the entire set of HOLC security rating polygons for a given urban area, and then compare these anomalies between cities in a quantitative manner.

We also estimate average percent developed impervious surface land cover [34] and tree canopy cover [35] within each HOLC polygon (Figure 2e,g) in each urban area as derived from the National Land Cover Database (NLCD) 2011 [36]. NLCD tree canopy percent is a 30 m raster dataset covering the coterminous United States, providing continuous percent tree canopy estimates derived from multi-spectral Landsat imagery for each 30 m pixel. NLCD imperviousness reports the percentage of urban developed surfaces that is impervious over every 30 m pixel in the coterminous United States and beyond. These estimates of underlying land use and overlying tree canopy may not sum to 100 percent, as tree canopy can exist over all land use types within a HOLC polygon and not all land use is necessarily impervious.

To compare δLST variations within and among HOLC security ratings between cities, we then average the estimated δLST by HOLC security rating category within each city. This binning by HOLC category yields how δLST varies between HOLC security ratings within each city. We then binned the δLSTs for each city at the national scale (n = 108) and by US Census Bureau regions: Northeast (n = 26), South (n = 29), Midwest (n = 41), and Western (n = 12). To estimate the significance of mean temperature differences between the HOLC security ratings by region and nationally, we apply a post-hoc ANOVA multiple comparisons test known as Tukey's Honest Significant Differences (HSD) Test. Tukey's HSD test estimates differences among group sample means for statistical significance. This pairwise post-hoc ANOVA test determines the statistical significance of differences between the mean of all pairs of group means using a studentized range distribution.

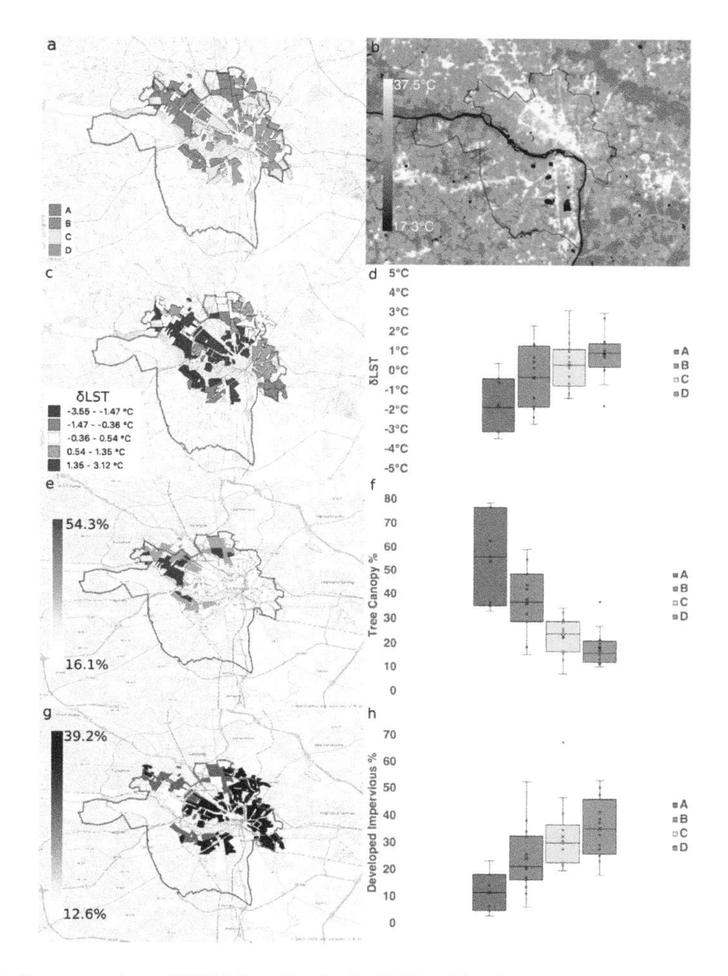

Figure 2. Demonstration of HOLC Security Grade δLST and land cover analysis for Richmond, VA (grey outline). (**a**) HOLC Polygons for Richmond, VA [29] (see Introduction text for explanations of HOLC security grade color designations), (**b**) LST map for Richmond, VA derived from Landsat 8 TIRS Band 10 imagery collected on 2 July 2016, and (**c**) Resulting δLSTs in HOLC polygons calculated as the anomaly of an individual HOLC polygon to the city-wide HOLC polygon average LST (see Equation (1)), (**d**) box–whisker plot of the δLSTs presented in (**c**) binned by HOLC security rating (see Introduction text for explanations of designations), (**e**) percent tree canopy from NLCD 2011 [36] averaged into HOLC polygons, (**f**) box–whisker plot of the δLSTs presented in (**e**) binned by HOLC security rating, (**g**) percent developed impervious surface from NLCD 2011 [36] averaged into HOLC polygons, (**h**) box–whisker plot of the average imperviousness presented in (**g**) binned by HOLC security rating.

3. Results

Our LST maps were generated from Landsat 8 acquisitions that satisfied a < 10 percent scene cloud coverage threshold collected from 3 June 2014 to 25 August 2017 (Supplementary Materials Table S1). These mostly sunny days provide the best conditions for Landsat 8 to reliably capture a strong LST pattern in urban areas [32]. Approximately 40 percent of the Landsat 8 imagery was collected during the 2016 northern hemisphere summer, while ~10 percent of the imagery was collected during the 2014 northern hemisphere summer. Regression tests reveal an insignificant relationship between the day that the imagery was collected and the resulting δLST patterns (Supplementary Materials Table S1).

Our analysis reveals three major trends that help to address our research questions. First, LST differences across the cities follow a non-uniform distribution of differences, suggesting that historical redlining policies are reflected in present-day intra-urban heat differentially (Supplementary Materials Table S1). Notable, intra-city δLST differences between areas given "D" and "A" HOLC security ratings range between +7.1 °C (Portland, OR) to −1.5 °C (Joliet, IL, USA), with ~94% of urban areas included in this study showing warmer present-day LSTs in their "D"-rated areas relative to their "A"-rated areas (Supplementary Materials Table S1). While Portland (OR) and Denver (CO) had the greatest "D" to "A" security rating differences within a city, the warmest δLST temperatures in formerly redlined areas relative to the city-wide average LST were identified in Chattanooga (TN, 3.3 °C) and Baltimore (MD, 3.2 °C). These cities were in contrast to formerly redlined areas that displayed, on average, cooler surface temperatures than their non-redlined counterparts (e.g., Joliet, IL, USA and Lima, OH, USA), a consistent pattern in several cities across the Midwest (Supplementary Materials Table S1). Patterns of relatively pronounced or muted δLST are underscored by attendant patterns of land use type and cover within the same HOLC security rating polygons, whereby the urban areas with the highest D-A difference and largest δLST in D-rated polygons show considerable HOLC rating-specific trends in average tree canopy and developed impervious surface percentages as compared to the Midwestern cities that exhibit cooler-than-average δLST patterns in their D-rated areas (Supplementary Materials Figure S1). The coolest δLST temperatures in areas assigned "A" HOLC security ratings relative to the city-wide average LST were identified in Birmingham (AL, −4.7 °C) and Roanoke (VA, −4.5 °C).

Regional aggregation of the city-specific trends reveals that average δLST differences between HOLC security rating categories exhibit a pattern of incremental warming relative to worsening HOLC security rating (Figure 3b–e). However, the magnitude of the δLST differences varies considerably by region, with the Midwest ($n = 41$) showing more muted δLST differences than the Southeast ($n = 29$) and West ($n = 12$), respectively (Figure 3b–e). Honest Significant Difference tests on urban areas at the regional scale reveal that the greatest δLST differences exist between "A" and "D" HOLC security rating areas across US regions, with "D"-rated areas progressively warmer than each subsequent rating in the present day. These amplified differences in the West and Southeast, as well as the relatively muted response in the Midwest (Figure 3b–e), are attended by similar differences in underlying percent land use cover (Figure 4b–e), and especially apparent in the available tree canopy (Figure 5b–e) for the areas assigned "A" HOLC security ratings.

A third trend that is consistent in a national-scale aggregation of δLSTs in these cities is the finding that "D"-rated areas are now on average 2.6 °C warmer than "A"-rated areas (Figure 3a). Each HOLC security rating category warms systematically relative to the more favorable neighbor security rating category (Figure 3a). Honest Significant Difference tests reveal that areas given "D" HOLC security ratings are significantly warmer than all of the other HOLC security rating categories at the national scale, in progressively larger magnitudes. These LST differences are underscored by similar, but opposing, national-scale patterns in underlying land use and tree canopy within the same redlined cities (Figures 4a and 5a), showing that areas assigned a "hazardous" HOLC security rating in US cities exhibit quantitatively less coverage by tree canopy and more coverage by impervious surfaces in the present decade [35,36].

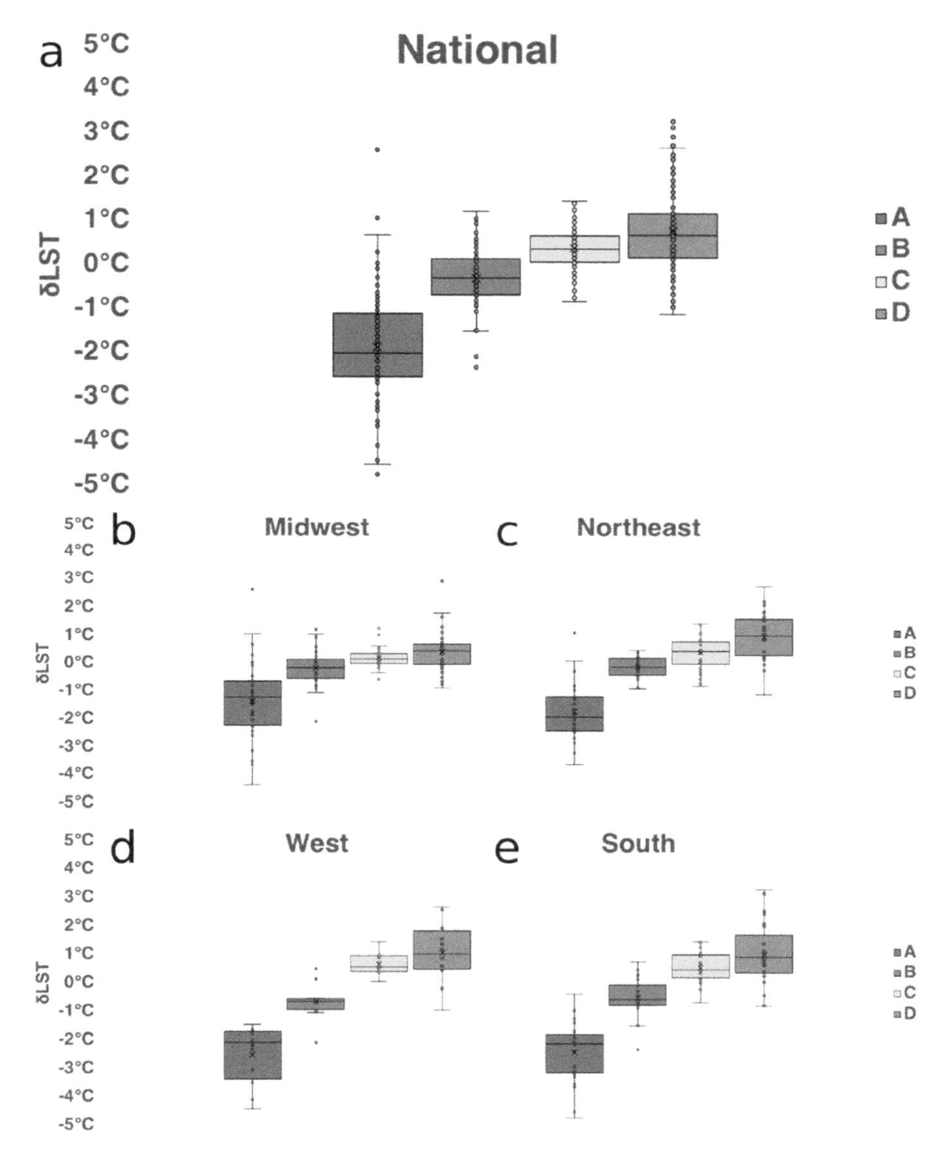

Figure 3. (**a**) National-scale Land Surface Temperature Anomalies by HOLC security rating (Green, "Best," A; Blue, "Still Desirable," B; Yellow, "Definitely Declining," C; Red, "Hazardous," D) (**b**) same as (**a**), but for the Midwest region; (**c**) same as (**b**), but for Northeast region; (**d**) same as (**b**), but for West region; (**e**) same as (**b**), but for South region.

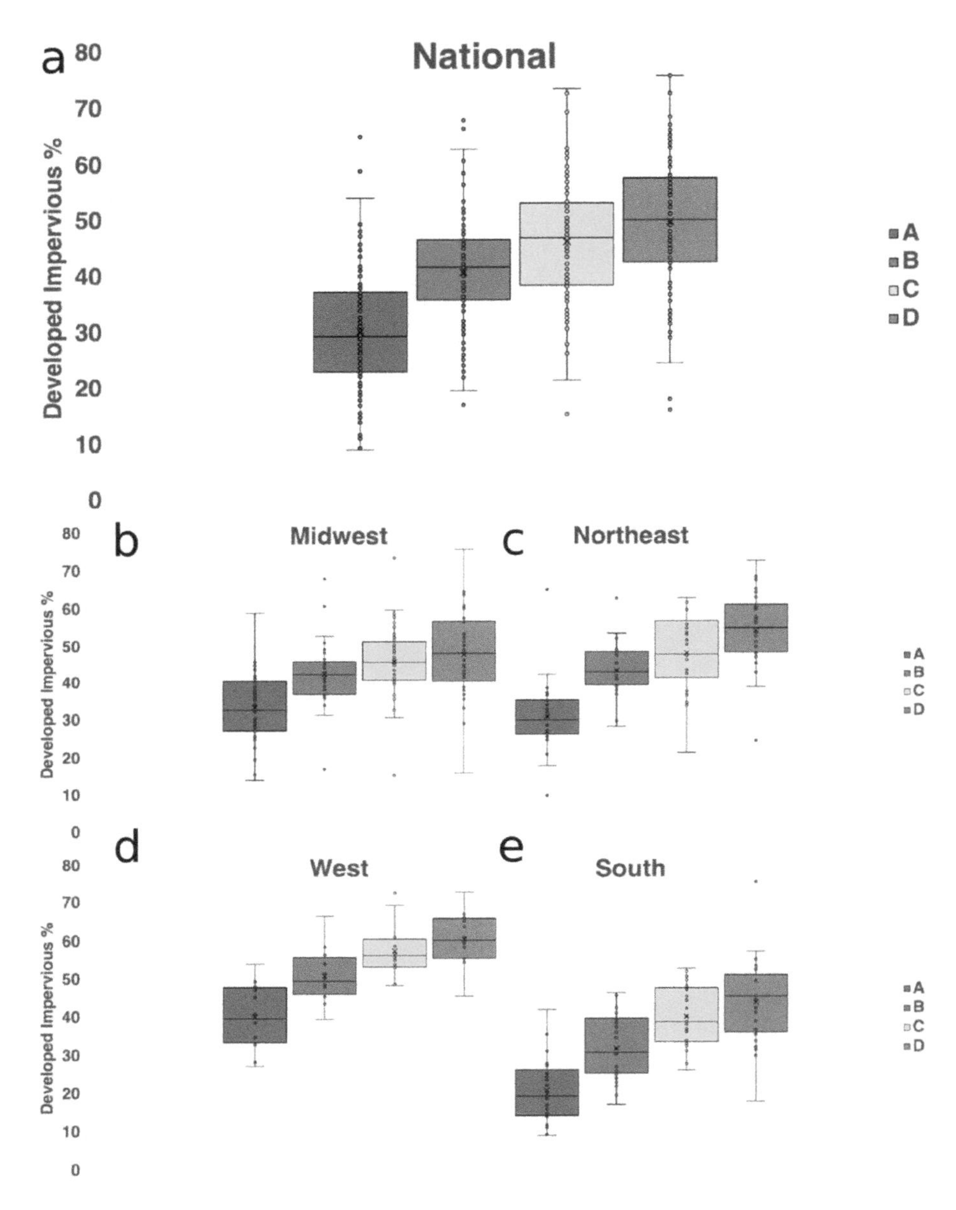

Figure 4. (**a**) National-scale averages of underlying percent developed impervious surface [36] by HOLC security rating (Green, "Best," A; Blue, "Still Desirable," B; Yellow, "Definitely Declining," C; Red, "Hazardous," D), (**b**) same as (**a**), but for the Midwest region; (**c**) same as (**b**), but for Northeast region; (**d**) same as (**b**), but for West region; (**e**) same as (**b**), but for South region.

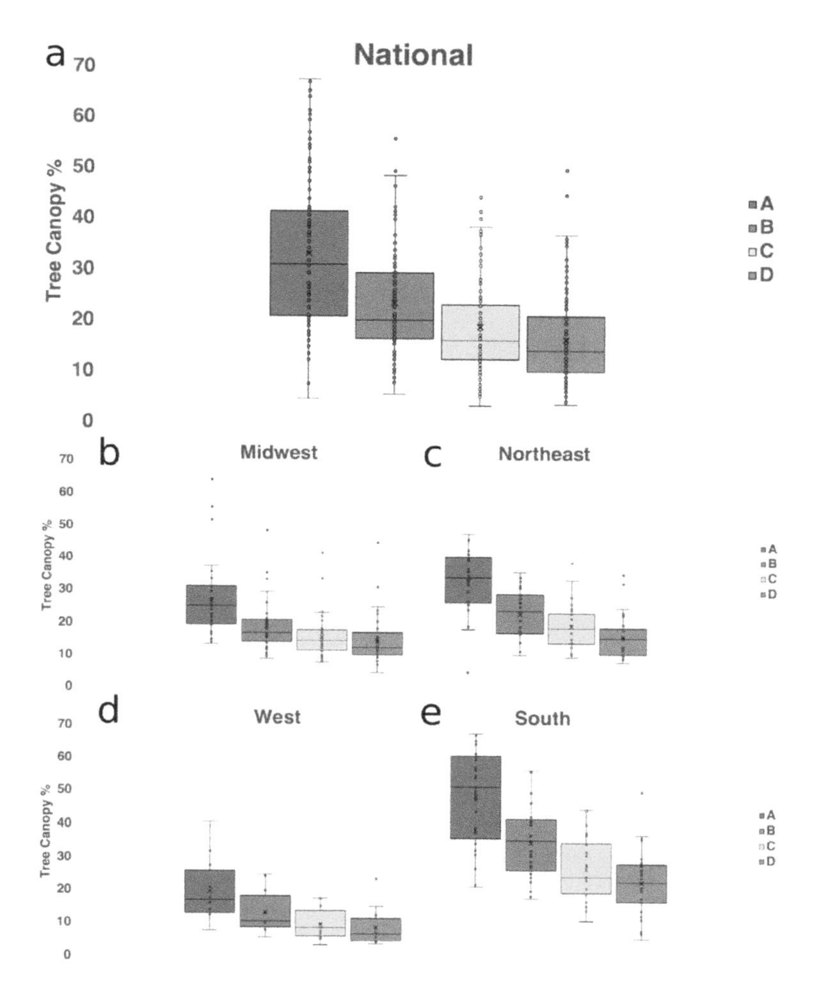

Figure 5. (**a**) National-scale averages of percent tree canopy [35,36] by HOLC security rating (Green, "Best," A; Blue, "Still Desirable," B; Yellow, "Definitely Declining," C; Red, "Hazardous," D), (**b**) same as (**a**), but for the Midwest region; (**c**) same as (**b**), but for Northeast region; (**d**) same as (**b**), but for West region; (**e**) same as (**b**), but for South region.

4. Discussion and Conclusions

We sought to understand the extent to which historic policies of redlining help to explain current patterns of intra-urban heat and the extent to which these patterns were consistent across US cities. Questions about the increasing economic inequality in US society motivated our inquiry and suggest several patterns related to historical federal housing policies and which communities experience the hottest areas of a city in the present day. Most notably, the consistency of greater temperature in formerly redlined areas across the vast majority (94%) of the cities included in this study indicates that current maps of intra-urban heat echo the legacy of past planning policies. While earlier studies document the lack of present-day services for and lower income of communities living in formerly redlined areas, this analysis presents an argument for understanding how global climate change will further exacerbate existing, historically-codified inequities in the US. We highlight three important dimensions of our findings—built environment, policies, and current inequities—as they relate to implications of these results.

First, our findings corroborate earlier studies that describe consistent patterns between the lack of tree canopy and historically underserved urban areas, at the national and regional scales (Figures 4 and 5). The prevalence of impervious surfaces as opposed to tree canopy points to the fact that green spaces have been consistently more abundant in wealthier and majority White-identifying neighborhoods [26]. At the same time, intra-urban heat is not only affected by tree cover, since the use of different materials within varying urban typologies also amplifies temperatures [19,37]. Two features of the urban landscape—roadways and large building complexes—are well known to transform solar radiation into heat. These landscape features absorb the energy-filled short-wave radiation coming from the sun, and re-emit long-wave radiation during the diurnal heating-cooling process. As a result, large roadways and building complexes gain heat during the day and, as the evening cools ambient temperatures, the retained heat is released back into the neighborhoods, which is captured by overhead satellite sensors. These evening temperatures are precisely the factors that can exacerbate excess mortality and morbidity [38].

An earlier body of evidence from the regional studies and economics literature makes the connection between federal programs that provided incentives for major roadway and building construction projects and the fact that many of these occurred in the lowest income neighborhoods of cities [39–41]. In fact, the 1950s were an important decade for the creation of major roadways across the US, and many redlined neighborhoods were transformed and divided by road and highway infrastructure projects [42]. These changes came at a time when intra-urban heat was not recognized as a major public health hazard, and yet, given the well-known heat-absorbing capacity of asphalt and concrete [43], the selection of these materials may underlie the differences revealed in these results.

Similarly, throughout the mid-1900s large building complexes, including housing complexes, industries, and university campuses, often subsidized by the federal government, were also placed in redlined areas, largely due to the inexpensive land, and current population of largely lower income and communities of color [44]. From the 1940s through the 1970s, large buildings were made of high-density materials, such as cinder block and brick, which retain heat, and maintain high temperatures through the night [45,46]. Many of these buildings still stand, and the LST maps investigated here partially describe the thermal signature of these buildings. Areas that were in non-redlined areas, often built of other materials but also dispersed across a more natural, maintained landscape, which allows for greater circulation of air [47,48] are hence the cooler neighborhoods registered by satellites.

Second, differences in implementing policies and landscapes may help to explain the variation of temperatures across different regions of the US. The cities of Portland (OR), Denver (CO), and Minneapolis (MN), for example, notably reflect the largest differences between the formerly redlined areas and their non-redlined counterparts (Supplementary Materials Table S1). We can speculate that the redlined areas of all three cities are currently located in areas with extensive physical infrastructure, including housing complexes, railway terminals, industrial or manufacturing sites, and/or adjacent to major business centers. The presence of these current day land uses may suggest a relationship between formerly inexpensive land and large-scale development. These results, when combined with more pernicious modern-day policies that support development of high-asphalt and low tree canopy areas such as massive shopping complexes that contain large surface parking lots, are further strengthened. In Portland (OR), for example, decades of development code allowed for multifamily complexes to cover 100% of the lot area with no provisions for greening. Only recently, and due to extensive support from local researchers and community organizations did the city evaluate earlier asphalt-driving policy to require 85% lot coverage and green spaces [49,50]. Such reversals of policy are the forms of planning that can help to reverse decades of amplifying temperatures in areas that have historically been underserved. Denver and Minneapolis are also making strides, though without further understanding about the historic and present-day drivers that generate these asphalt-rich and tree canopy-poor land uses on intra-urban heat, and local communities, progress will be slow.

In addition, the coupling of landscape and historic designs of urban development in these cities may also play a role in helping to explain differences across the country. Portland, like Minneapolis, are in landscapes where tree canopy is relatively easy to sustain. Unlike arid and drought-prone areas, where planting trees can require extensive maintenance, the warm, sunny summers and wet/snowy winters of Portland, Minneapolis and other cities of the Northwest and Midwest, provide ideal conditions for expanding an urban forest, which can, in turn, reduce surface temperatures of a neighborhood. Tree planting efforts often took place as part of urban development projects in the early and mid-1900s, and were used as a way to mark special designations [42]. Similarly, metropolitan areas that conform to the concentric zone model (for example, places like Chicago, Los Angeles, and Philadelphia) tend to be larger and more densely populated metros, often with a higher degree of both affluence and inequality, a larger African American population, and a greater share of population in the suburbs. In the remaining metropolitan areas, there is greater integration between the affluent and the poor [44]. In these places, such as Seattle (WA), Charleston (WV), and Birmingham (AL), the rich are concentrated in the urban core, where redlining and tree planting efforts coincide.

Finally, indicators of and/or higher intra-urban LSTs have been shown to correlate with higher summertime energy use [51,52], and excess mortality and morbidity [20,53,54]. The fact that residents living in formerly redlined areas may face higher financial burdens due to higher energy and more frequent health bills further exacerbates the long-term and historical inequities of present and future climate change. As the results from earlier studies have documented income inequality between formerly redlined areas another other parts of US cities, we recognize that hotter areas will amplify these current inequities. Such historic income inequality leads to income segregation because higher incomes, which are further supported by past and current housing policy, allow certain households to sort themselves according to their preferences—and control local political processes that continue exclusion [55]. Other explanatory factors of these patterns, though too many for the current study and setting the stage for future studies, include disinvestment in urban areas, suburban investment and land use patterns, and the practices generally of government and the underwriting industry [39,56].

To our knowledge, this is the first study to link a historical federal housing policy to the creation (or at least the exacerbation) of a climate stressor and potential variability in resident exposure to it. While redlining most likely did not create the microenvironments that mediate LSTs relative to the rest of the urban environment, our findings suggest a strong and significant likelihood of the cauterization of current day exposure to the hottest parts of a city. While patterns of who experiences the most exposure to intra-urban heat may change as a result of (green) gentrification, which many formerly redlined neighborhoods are undergoing (e.g., wealthier communities can afford to green and change the physical landscape, and, over time, cool the hottest areas of a city), we observe consistent patterns that can be inferred as in part due to the creation of HOLC maps [23]. Future studies will need to describe the mechanisms by with planning practices—past and present—are likely to amplify the effects of climate change on historically underserved communities and communities of color.

While a growing body of evidence describes the intra-urban variation of temperatures due to characteristics of the built environment, few have asked why we observe a pattern of historically-marginalized communities living in the hottest areas. Here we have presented results from an analysis of 108 US cities that aimed to examine the role of historic "redlining" policies in mediating exposure to intra-urban heat. We found that in nearly all cases, those neighborhoods located in formerly redlined areas—that remain predominantly lower income and communities of color—are at present hotter than their non-redlined counterparts. Although the extent of differences in temperatures varies by region, the preponderance of evidence establishes that those experiencing the greatest exposure to present and potentially future extreme heat are living in neighborhoods with the least social and ecosystem services historically.

As more and more communities race to develop plans to react to and adapt to worsening extreme heat and its attendant effects on human health [57], a research agenda focused on developing place-specific, heat-mitigating urban designs and interventions [58–61] will be critical toward not only alleviating heat disparity but ensuring that the urban forms and policies that gave rise to these inequities in our past (like redlining) are recognized and altogether avoided. Furthermore, crafting climate equity-centered policies that recognize decades of disproportionate exposure to environmental stressors can help any new discoveries in urban design get implemented with focus and rapidity.

Supplementary Materials: The following are available online at http://www.mdpi.com/2225-1154/8/1/12/s1, Figure S1: Comparison of urban areas of relatively large or small differences in LSTs between HOLC grades, Table S1: Urban area-specific results from our LST analysis.

Author Contributions: J.S.H. conceived the project and coordinated the analysis, advised interpretation, and contributed to the manuscript, created figures, and coordinated the responses to reviewers. V.S. provided major contributions to the manuscript including context, interpretation, editorialization, and literature review and references. N.P. performed HOLC map and satellite imagery download and spatial analyses. All authors have read and agreed to the published version of the manuscript.

Funding: This research involved no external funding.

Acknowledgments: J.S.H. thanks the NOAA Office of Education Environmental Literacy Program, the Virginia Academy of Science, Groundwork RVA, Virginia Commonwealth University SustainLab, University of Richmond Spatial Analysis and Digital Scholarship Labs, and the City of Richmond Sustainability Office. J.S.H. and V.S. acknowledge support from the NOAA Climate Program Office, and U.S. Forest Service's National Urban and Community Forestry Challenge Grants Program (No. 17-DG-11132544-014). N.P. acknowledges the work study program at the Virginia Commonwealth University. The authors thank four anonymous reviewers for their thoughtful and thorough consideration.

Conflicts of Interest: The authors declare no conflict of interest.

References

1. Wong, K.V.; Paddon, A.; Jimenez, A. Review of World Urban Heat Islands: Many Linked to Increased Mortality. *Energy Resour. Technol.* **2013**, *135*, 022101. [CrossRef]
2. Poumadère, M.; Mays, C.; Le Mer, S.; Blong, R. The 2003 Heat Wave in France: Dangerous Climate Change Here and Now. *Risk Anal.* **2005**, *25*, 1483–1494. [CrossRef]
3. Borden, K.A.; Cutter, S.L. Spatial patterns of natural hazards mortality in the United States. *Int. J. Health Geogr.* **2008**, *7*, 64. [CrossRef]
4. Hess, J.J.; Eidson, M.; Tlumak, J.E.; Raab, K.K.; George, L. An evidence-based public health approach to climate change adaptation. *Environ. Health Perspect.* **2014**, *122*, 1177–1186. [CrossRef]
5. Uejio, C.K.; Wilhelmi, O.V.; Golden, J.S.; Mills, D.M.; Gulino, S.P.; Samenow, J.P. Intra-urban societal vulnerability to extreme heat: The role of heat exposure and the built environment, socioeconomics, and neighborhood stability. *Health Place* **2011**, *17*, 498–507. [CrossRef]
6. Habeeb, D.; Vargo, J.; Stone, B. Rising heat wave trends in large US cities. *Nat. Hazards* **2015**, *76*, 1651–1665. [CrossRef]
7. Wang, Y.; Wang, A.; Zhai, J.; Tao, H.; Jiang, T.; Su, B.; Yang, J.; Wang, G.; Liu, Q.; Gao, C.; et al. Tens of thousands additional deaths annually in cities of China between 1.5 °C and 2.0 °C warming. *Nat. Commun.* **2019**, *10*, 3376. [CrossRef] [PubMed]
8. Meehl, G.A.; Tebaldi, C. More Intense, More Frequent, and Longer Lasting Heat Waves in the 21st Century. *Science* **2004**, *305*, 994. [CrossRef] [PubMed]
9. Santamouris, M. Analyzing the heat island magnitude and characteristics in one hundred Asian and Australian cities and regions. *Sci. Total Environ.* **2015**, *512–513*, 582–598. [CrossRef]
10. Lo, Y.T.E.; Mitchell, D.M.; Gasparrini, A.; Vicedo-Cabrera, A.M.; Ebi, K.L.; Frumhoff, P.C.; Millar, R.J.; Roberts, W.; Sera, F.; Sparrow, S.; et al. Increasing mitigation ambition to meet the Paris Agreement's temperature goal avoids substantial heat-related mortality in U.S. cities. *Sci. Adv.* **2019**, *5*, eaau4373. [CrossRef] [PubMed]
11. Voelkel, J.; Shandas, V.; Haggerty, B. Developing High-Resolution Descriptions of Urban Heat Islands: A Public Health Imperative. *Prev. Chronic Dis.* **2016**, *13*, E129. [CrossRef]
12. Ziter, C.D.; Pedersen, E.J.; Kucharik, C.J.; Turner, M.G. Scale-dependent interactions between tree canopy cover and impervious surfaces reduce daytime urban heat during summer. *Proc. Natl. Acad. Sci. USA* **2019**, *116*, 7575. [CrossRef]

13. Shandas, V.; Voelkel, J.; Williams, J.; Hoffman, J. Integrating Satellite and Ground Measurements for Predicting Locations of Extreme Urban Heat. *Climate* **2019**, *7*, 5. [CrossRef]
14. Li, X.; Zhou, Y.; Asrar, G.R.; Imhoff, M.; Li, X. The surface urban heat island response to urban expansion: A panel analysis for the conterminous United States. *Sci. Total Environ.* **2017**, *605–606*, 426–435. [CrossRef]
15. Li, Y.-Y.; Zhang, H.; Kainz, W. Monitoring patterns of urban heat islands of the fast-growing Shanghai metropolis, China: Using time-series of Landsat TM/ETM+ data. *Int. J. Appl. Earth Obs. Geoinf.* **2012**, *19*, 127–138. [CrossRef]
16. Davis, A.Y.; Jung, J.; Pijanowski, B.C.; Minor, E.S. Combined vegetation volume and "greenness" affect urban air temperature. *Appl. Geogr.* **2016**, *71*, 106–114. [CrossRef]
17. Jun, M.-J.; Kim, J.-I.; Kim, H.-J.; Yeo, C.-H.; Hyun, J.-Y. Effects of Two Urban Development Strategies on Changes in the Land Surface Temperature: Infill versus Suburban New Town Development. *J. Urban Plann. Dev.* **2017**, *143*, 04017010. [CrossRef]
18. Aram, F.; Higueras García, E.; Solgi, E.; Mansournia, S. Urban green space cooling effect in cities. *Heliyon* **2019**, *5*, e01339. [CrossRef]
19. Hamstead, Z.A.; Kremer, P.; Larondelle, N.; McPhearson, T.; Haase, D. Classification of the heterogeneous structure of urban landscapes (STURLA) as an indicator of landscape function applied to surface temperature in New York City. *Ecol. Indic.* **2016**, *70*, 574–585. [CrossRef]
20. Madrigano, J.; Ito, K.; Johnson, S.; Kinney, P.L.; Matte, T. A Case-Only Study of Vulnerability to Heat Wave–Related Mortality in New York City (2000–2011). *Environ. Health Perspect.* **2015**, *123*, 672–678. [CrossRef]
21. Voelkel, J.; Hellman, D.; Sakuma, R.; Shandas, V. Assessing Vulnerability to Urban Heat: A Study of Disproportionate Heat Exposure and Access to Refuge by Socio-Demographic Status in Portland, Oregon. *IJERPH* **2018**, *15*, 640. [CrossRef]
22. Whitman, S.; Good, G.; Donoghue, E.R.; Benbow, N.; Shou, W.; Mou, S. Mortality in Chicago attributed to the July 1995 heat wave. *Am. J. Public Health* **1997**, *87*, 1515–1518. [CrossRef]
23. Aaronson, D.; Hartley, D.; Mazumder, B. The Effects of the 1930s HOLC "Redlining" Maps. In *Federal Reserve Bank of Chicago Working Paper No. 2017-12*; Federal Reserve Bank of Chicago: Chicago, IL, USA, 2017; pp. 1–102.
24. Mitchell, B.; Franco, J. *HOLC "Redlining" Maps: The Persistent Structure of Segregation and Economic Inequality*; National Community Reinvestment Coalition: Washington, DC, USA, 2018; pp. 1–29.
25. Lipsitz, G. *How Racism Takes Place*; Temple University Press: Philadelphia, PA, USA, 2011.
26. Nowak, D.J.; Greenfield, E.J. Declining urban and community tree cover in the United States. *Urban For. Urban Green.* **2018**, *32*, 32–55. [CrossRef]
27. Schwarz, K.; Fragkias, M.; Boone, C.G.; Zhou, W.; McHale, M.; Grove, J.M.; O'Neil-Dunne, J.; McFadden, J.P.; Buckley, G.L.; Childers, D.; et al. Trees Grow on Money: Urban Tree Canopy Cover and Environmental Justice. *PLoS ONE* **2015**, *10*, e0122051. [CrossRef]
28. Nardone, A.; Thakur, N.; Balmes, J.R. Historic Redlining and Asthma Exacerbations across Eight Cities of California: A Foray into How Historic Maps Are Associated with Asthma Risk. *Am. J. Resp. Crit. Care Med* **2019**, *199*, A7054.
29. Nelson, R.K.; Winling, L.; Marciano, R.; Connolly, N. Mapping Inequality: Redlining in New Deal America. Available online: https://dsl.richmond.edu/panorama/redlining/ (accessed on 9 October 2017).
30. Avdan, U.; Jovanovska, G. Algorithm for Automated Mapping of Land Surface Temperature Using LANDSAT 8 Satellite Data. *J. Sens.* **2016**, *2016*, 1–8. [CrossRef]
31. Cook, M.; Schott, J.; Mandel, J.; Raqueno, N. Development of an Operational Calibration Methodology for the Landsat Thermal Data Archive and Initial Testing of the Atmospheric Compensation Component of a Land Surface Temperature (LST) Product from the Archive. *Remote Sens.* **2014**, *6*, 11244–11266. [CrossRef]
32. Sheng, L.; Tang, X.; You, H.; Gu, Q.; Hu, H. Comparison of the urban heat island intensity quantified by using air temperature and Landsat land surface temperature in Hangzhou, China. *Ecol. Indic.* **2017**, *72*, 738–746. [CrossRef]
33. White-Newsome, J.L.; Brines, S.J.; Brown, D.G.; Dvonch, J.T.; Gronlund, C.J.; Zhang, K.; Oswald, E.M.; O'Neill, M.S. Validating Satellite-Derived Land Surface Temperature with *in situ* Measurements: A Public Health Perspective. *Environ. Health Perspect.* **2013**, *121*, 925–931. [CrossRef]

34. Yang, L.; Huang, C.; Homer, C.G.; Wylie, B.K.; Coan, M.J. An approach for mapping large-area impervious surfaces: Synergistic use of Landsat-7 ETM+ and high spatial resolution imagery. *Can. J. Remote Sens.* **2014**, *29*, 230–240. [CrossRef]
35. Coulston, J.W.; Moisen, G.G.; Wilson, B.T.; Finco, M.V.; Cohen, W.B.; Brewer, C.K. Modeling percent tree canopy cover: A pilot study. *Photogramm. Eng. Remote Sens.* **2012**, *78*, 715–727. [CrossRef]
36. Homer, C.G.; Dewitz, J.; Yang, L.; Jin, S.; Danielson, P.; Xian, G.Z.; Coulston, J.; Herold, N.; Wickham, J.; Megown, K. Completion of the 2011 National Land Cover Database for the conterminous United States—Representing a decade of land cover change information. *Photogramm. Eng. Remote Sens.* **2015**, *81*, 345–354.
37. Stewart, I.D.; Oke, T.R. Local Climate Zones for Urban Temperature Studies. *Bull. Am. Meteor. Soc.* **2012**, *93*, 1879–1900. [CrossRef]
38. Murage, P.; Hajat, S.; Kovats, R.S. Effect of night-time temperatures on cause and age-specific mortality in London. *Environ. Epidemiol.* **2017**, *1*. [CrossRef]
39. Hirsch, A. *Making the Second Ghetto: Race and Housing in Chicago, 1940–1960*; Cambridge University Press: New York, NY, USA, 1983.
40. Teaford, J.C. *The Rough Road to Renaissance: Urban Revitalization in America, 1940–1985*; Johns Hopkins University Press: Baltimore, MD, USA, 1990.
41. Sugrue, T.J. *The Origins of the Urban Crisis: Race and Inequality in Postwar Detroit*; Princeton University Press: Princeton, NJ, USA, 1996.
42. DiMento, J.F.C. Stent (or Dagger?) in the Heart of Town: Urban Freeways in Syracuse, 1944–1967. *J. Plan. Hist.* **2009**, *8*, 133–161. [CrossRef]
43. Swaid, H. Numerical investigation into the influence of geometry and construction materials on urban street climate. *Phys. Geogr.* **2013**, *14*, 342–358. [CrossRef]
44. Zuk, M.; Bierbaum, A.H.; Chapple, K.; Gorska, K.; Loukaitou-Sideris, A. Gentrification, Displacement, and the Role of Public Investment. *J. Plan. Lit.* **2017**, *33*, 31–44. [CrossRef]
45. Kim, H.H. Urban heat island. *Int. J. Remote Sens.* **1992**, *13*, 2319–2336. [CrossRef]
46. Hall, J.P. *The Early Developmental History of Concrete Block in America*; Ball State University Library: Muncie, IN, USA, 2009.
47. Howard, B.; Parshall, L.; Thompson, J.; Hammer, S.; Dickinson, J.; Modi, V. Spatial distribution of urban building energy consumption by end use. *Energy Build.* **2012**, *45*, 141–151. [CrossRef]
48. Barrington-Leigh, C.; Millard-Ball, A. A century of sprawl in the United States. *Proc. Natl. Acad. Sci. USA* **2015**, *112*, 8244–8249. [CrossRef]
49. Anderson, M. In Mid-Density Zones, Portland Has a Choice: Garages or Low Prices? Available online: https://www.sightline.org/2019/10/02/in-mid-density-zones-portland-has-a-choice-garages-or-low-prices/ (accessed on 14 October 2019).
50. Better Housing by Design. Available online: https://www.portlandoregon.gov/bps/71903 (accessed on 14 October 2019).
51. Lowe, S.A. An energy and mortality impact assessment of the urban heat island in the US. *Environ. Impact Assess. Rev.* **2016**, *56*, 139–144. [CrossRef]
52. Li, X.; Zhou, Y.; Yu, S.; Jia, G.; Li, H.; Li, W. Urban heat island impacts on building energy consumption: A review of approaches and findings. *Energy* **2019**, *174*, 407–419. [CrossRef]
53. Chuang, W.-C.; Gober, P. Predicting Hospitalization for Heat-Related Illness at the Census-Tract Level: Accuracy of a Generic Heat Vulnerability Index in Phoenix, Arizona (USA). *Environ. Health Perspect.* **2015**, *123*, 606–612. [CrossRef]
54. Eisenman, D.P.; Wilhalme, H.; Tseng, C.-H.; Chester, M.; English, P.; Pincetl, S.; Fraser, A.; Vangala, S.; Dhaliwal, S.K. Heat Death Associations with the built environment, social vulnerability and their interactions with rising temperature. *Health Place* **2016**, *41*, 89–99. [CrossRef]
55. Reardon, S.F.; Bischoff, K. Income Inequality and Income Segregation. *Am. J. Sociol.* **2011**, *116*, 1092–1153. [CrossRef]
56. Levy, D.K.; McDade, Z.; Dumlao, K. *Effects from Living in Mixed-Income Communities for Low-Income Families*; Urban Institute: Washington, DC, USA, 2011; p. 34.
57. Martinez, G.S.; Linares, C.; Ayuso, A.; Kendrovski, V.; Boeckmann, M.; Diaz, J. Heat-health action plans in Europe: Challenges ahead and how to tackle them. *Environ. Res.* **2019**, *176*, 108548. [CrossRef]

58. Makido, Y.; Hellman, D.; Shandas, V. Nature-Based Designs to Mitigate Urban Heat: The Efficacy of Green Infrastructure Treatments in Portland, Oregon. *Atmosphere* **2019**, *10*, 282. [CrossRef]
59. Hatvani-Kovacs, G.; Belusko, M.; Pockett, J.; Boland, J. Heat stress-resistant building design in the Australian context. *Energy Build.* **2018**, *158*, 290–299. [CrossRef]
60. Alam, M.; Sanjayan, J.; Zou, P.X.W. Chapter Eleven—Balancing Energy Efficiency and Heat Wave Resilience in Building Design. In *Climate Adaptation Engineering*; Bastidas-Arteaga, E., Stewar, M.G., Eds.; Butterworth-Heinemann: Oxford, UK, 2019; pp. 329–349.
61. He, B.-J. Towards the next generation of green building for urban heat island mitigation: Zero UHI impact building. *Sustain. Cities Soc.* **2019**, *50*, 101647. [CrossRef]

Article

Retrospective Analysis of Summer Temperature Anomalies with the Use of Precipitation and Evapotranspiration Rates

Andri Pyrgou [1], Mattheos Santamouris [2,*], Iro Livada [3] and Constantinos Cartalis [3]

1 Department of Civil Aviation, Pindarou 27 str., 1429 Nicosia, Cyprus
2 Anita Lawrence Chair High Performance Architecture, Faculty of Built Environment, University of New South Wales, Sydney 2033, Australia
3 Department of Environmental Physics, National and Kapodistrian University of Athens, 15784 Athens, Greece
* Correspondence: m.santamouris@unsw.edu.au

Received: 5 July 2019; Accepted: 28 August 2019; Published: 30 August 2019

Abstract: Drought and extreme temperatures forecasting is important for water management and the prevention of health risks, especially in a period of observed climatic change. A large precipitation deficit together with increased evapotranspiration rates in the preceding days contribute to exceptionally high temperature anomalies in the summer above the average local maximum temperature for each month. Using a retrospective approach, this study investigated droughts and extreme temperatures in the greater area of Nicosia, Cyprus and suggests a different approach in determining the lag period of summer temperature anomalies and precipitation. In addition, dry conditions defined with the use of the Standardized Precipitation-Evapotranspiration Index (SPEI) were associated with positive temperature anomalies at a percentage up to 33.7%. The compound effect of precipitation levels and evapotranspiration rates of the preceding days for the period 1988–2017 to summer temperature anomalies was demonstrated with significantly statistical R squared values up to 0.57. Furthermore, the cooling effect of precipitation was higher and prolonged longer in rural and suburban than urban areas, a fact that is directly related to the evaporation potential of the area in concern. Our work demonstrates the compound effect of precipitation levels and evapotranspiration rates of the preceding days to summer temperature anomalies.

Keywords: Mediterranean; semi-arid; drought; standardized precipitation evapotranspiration index (SPEI); climate warming; soil moisture

---

1. Introduction

Weather regimes drive climate change and influence temperature variation [1] and may persist from a few days to a few weeks. Weather regimes in Cyprus depend on mid-latitude flow dynamics, yet they are regulated by several external factors, such as dry soils [2,3] and sea-surface temperature anomalies [4,5] that subsequently affect the development and the duration of heat waves. The feasibility of prediction of extreme temperatures in the summer using numerical models largely rests on the variability of soil moisture, sea surface temperature, and heat fluxes [6]. Variations of surface temperature after a precipitation event in the summer suggest that, due to the wet ground, more energy is likely to go into evaporation at the expense of sensible heating [7,8]. Precipitation is also associated with clouds blocking the sun and provides less energy by further reducing the temperature [7,9].

Hirschi et al. [10] divided the European domain into two sectors based on the soil moisture variations: southeast Europe with transitional soil-moisture-limited evapotranspiration regime and central European characterized by a wet soil-moisture regime (energy-limited evapotranspiration regime) [10]. A strong relationship between soil-moisture deficit and summer hot extremes in southeast Europe was noted. Droughts and heatwaves have been shown to intensify and propagate via land–atmosphere feedbacks [3]. Fischer et al. [2] argued that a large precipitation deficit together with early vegetation green-up and strong positive radiative anomalies in the months preceding the extreme summer event contributed to an early and rapid loss of soil moisture [2], resulting in low latent cooling and increased temperatures. Soil moisture deficits induce higher temperatures of about 5–6 °C over the initially drier region [11]. Several studies have suggested that the variations of summer climate are regulated by the soil moisture-atmosphere interactions [12–14], because soil moisture acts as a storage component for precipitation and affects plant transpiration and photosynthesis with subsequent impacts on water, energy, and biogeochemical cycles [15]. Drivers of evapotranspiration vary with climate regimes, particularly in the transitional Mediterranean climate where soil moisture is limited. Regions may switch between energy-limited and soil moisture-limited evapotranspiration regimes through the year due to land cover [15]. McHugh et al. [16] studied soil moisture in semi-arid regions and showed that atmospheric moisture may significantly contribute to variations in soil water content. The study additionally showed that maximum respiration rates could arise in the early morning [16] when soils are warm enough to stimulate microbial activity and carbon cycling, and they still contain moisture trapped through water vapor adsorption [17]. In semi-arid climates, such as Cyprus, depletion of soil moisture occurs in the early summer (May–June), but other sources of soil moisture may be fog deposition, dew formation, and water vapor adsorption [17,18].

Liu et al. [19] articulated that soil moisture memory is approximately 2–3 months in mid-latitudes and that dry initial soil moisture anomalies lead to a decrease of precipitation and an increase of surface temperature in the subsequent months, resulting in an increase of droughts and hot and cold extremes [19]. Several drought indices have been adopted that investigate droughts using precipitation data or estimation of evaporative losses, which seriously alter the natural water availability [20]. In the case of limited precipitation, moisture stays only in the upper layers, whereas in abundance of rainfall, moisture reaches the lower layers and recharges the bedrock fractures. Increased atmospheric evaporative demand due to warming, solar radiation, humidity, and wind speed lead to further drying of the areas where precipitation reduces, resulting in droughts [20] as the drying of the surface is enhanced with water scarcity. Eliades et al. [21] studied the transpiration of *Pinus bruita* trees in the mountainous area of Cyprus for the years 2015 to 2017 and evidenced that high levels of rain and soil moisture in the preceding fall months can recharge the bedrock fractures, leading to higher transpiration in the early summer [21]. However, this mechanism also depends on leaf area and rooting depth. Enhancement of air moisture in the early summer may also be dependent on transpiration and the vegetation type. Extremely high temperatures and extended drought also affect the physiological processes in plants by regulating the stomatal openings, increasing the rate of photorespiration in leaves and irreversibly damaging leaves, leading to plant death [22].

Temperature anomalies are mostly affected by external climatic conditions, such as precipitation frequency, amount of precipitation, and synoptic weather conditions. The adaptation strategies should therefore aim to modify the vulnerability component by changing the adaptive capacity of a region to withstand extremely high or low temperatures. Vulnerability may change based on human capacity, social and cultural habits, governance of a region, and physical and biological parameters [23]. However, social vulnerability differs for heatwaves and drought for people who live in poorly constructed homes, older people, and those who work in hot conditions. Management options may accelerate adaptation to climatic variability because the response of each area to environmental conditions at any moment in time depends on the current state of the system and not on its past history of exposure to events.

In this study, the relationship between ambient air temperature anomalies in Cyprus and the preceding deficit in precipitation from the previous months was investigated via a retrospective approach and a solid statistical methodology for the period 1988–2017 (inclusive). This study used the cross-correlation analysis to determine the lag period of summer temperature anomalies and precipitation. The role of land albedo with soil moisture is important, thus we compared the lag period of three different areas under the same climatic conditions with contrasting land cover. Even though the land albedo was not quantified, the different characteristics of the urban and the rural layouts were obvious through the satellite images and the noteworthy results of the analysis. Moreover, this study examined the effect of summer precipitation and related relative risk factors for higher temperatures under drought conditions in each area; the analysis was comparatively applied in urban, suburban, or rural areas in order to identify how the built environment affects urban temperatures. Drought was defined with the use of the Standardized Precipitation-Evapotranspiration Index (SPEI) multi-scalar drought index that represents both the supply and the demand sides of the surface moisture balances by investigating the evapotranspiration rate of the preceding months for three nearby stations with different land-use in a semi-arid Mediterranean country. Results demonstrate the feasibility of the development of an operational early warning system and adaptation measures in southern Europe considering the vulnerability of the area to droughts.

2. Study Area and Datasets

Cyprus (Figure 1) is an island in the eastern basin of the Mediterranean Sea with an area of 9251 km^2. Cyprus has a hot summer Mediterranean climate and a hot semi-arid climate (in the northeastern part of island) according to Köppen climate classification signs Csa (Mediterranean hot summer climates) and BSh (Hot semi-arid climates) [24], with warm to hot dry summers and wet winters. The hot, dry summer lasting from May to September is affected by the low barometric centered in Southwest Asia, which contributes to the persistence of high temperatures and low precipitation levels.

Three meteorological stations were investigated: an urban station (35.17° N, 33.36° E) in the city center, a suburban station (35.15° N, 33.40° E), and a rural station (35.05° N, 33.54° E) at a distance of 21.3 km from the urban station (Figure 1). The urban, the suburban, and the rural stations are located at altitudes 160, 162, and 175 m above mean sea level, respectively (Figure 2). The maximum height of buildings is 24 m (six floors) at the urban area, 17 m (four floors) at the suburban area, and 8.3 m (two floors) at the rural area [25].

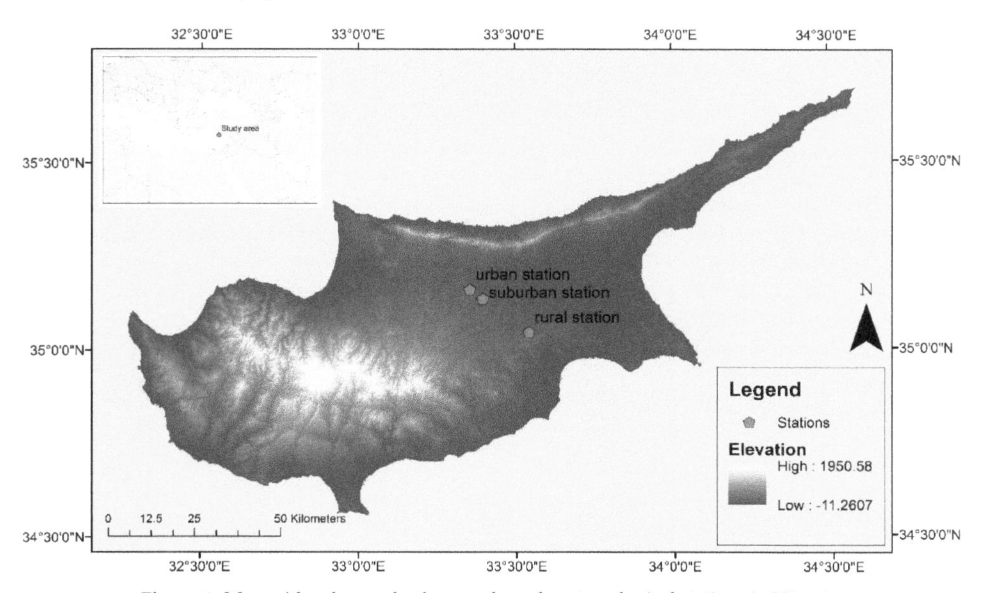

Figure 1. Map with urban, suburban, and rural meteorological stations in Nicosia.

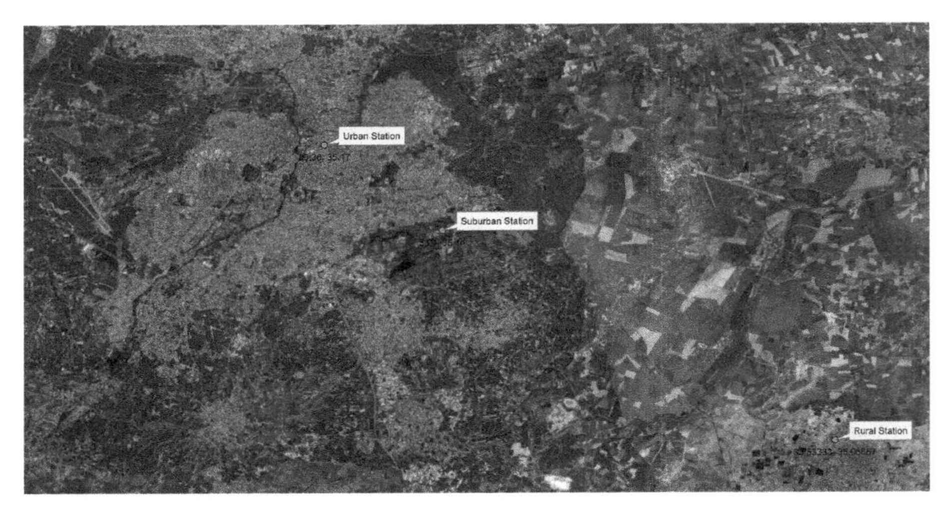

Figure 2. Geophysical map showing the landscape surrounding the three investigated areas (urban, suburban, rural).

The daily ambient air temperature (mean, maximum, and minimum) as well as the daily accumulated precipitation were obtained from the Meteorological Service of Cyprus for the period 1988–2017 (inclusive) [26]. Only the months April to September were chosen from the continuous dataset for further investigation. No outliers or missing data existed in the final dataset, ensuring normality and homogeneity of variance throughout the series. The mean ambient air temperatures for the months May to September were 27.6 °C, 27.1 °C, and 26.7 °C for the urban, the suburban, and the rural areas, respectively.

3. Methodology

3.1. Ambient Air Temperatures and Total Precipitation in the Urban, Suburban and Rural Areas

For the investigated years (1988–2017), a linear trend analysis was used to estimate the statistical significance of the slope (b) of trend lines and reveal specific patterns of the local climate of the monthly values of temperatures and precipitation for months April to September for the three stations. The t-test analysis was used to allow for comparisons with other studies that investigate increasing and decreasing trends of temperature, precipitation, and climatic abnormalities [27–30]. According to the t-test analysis (Table 1) for the regression lines, the maximum air temperatures showed a steady profile throughout the years (values less than 2.048 for $\alpha = 0.05$ and 28d.f), but the minimum and the mean temperatures showed a statistically increasing trend (values over than 2.048 for a = 0.05 and 28d.f).

Table 1. t-test (t_b values) for testing the significance of the slope of trend lines.

	Urban	Suburban	Rural
Tmax (°C)	0.745	1.344	1.452
Tmin (°C)	**5.576**	**8.813**	**8.284**
Tmean (°C)	**3.006**	**5.134**	**4.428**
Total precipitation (mm)	−1.712	1.158	0.112

The following table (Table 2) presents mean monthly maximum, minimum, and mean air temperatures and the total monthly precipitation for the three investigated areas (urban, suburban, and rural) for months May to September. The highest average monthly temperatures developed in

July, followed by August for all areas. Precipitation was the lowest in August with values close to zero. Moreover, the histograms (Figure 3a–i) show the distribution of these reference values.

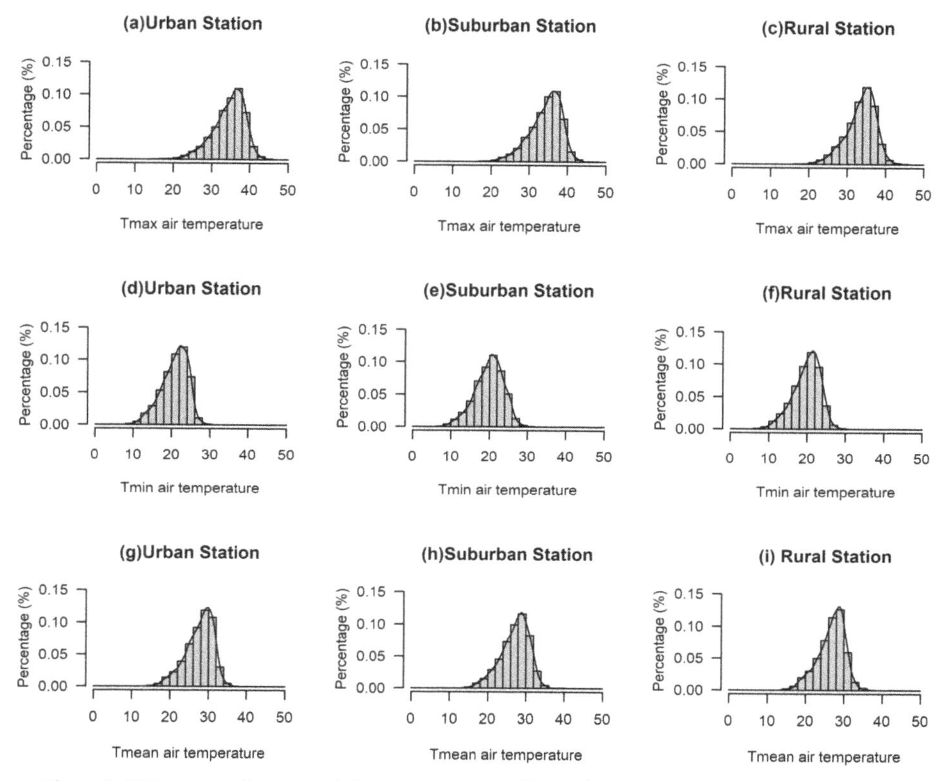

Figure 3. Histograms of summer daily measurements of Tmax (**a–c**), Tmin (**d–f**), and Tmean (**g–i**) for urban, suburban, and rural stations.

Figure 3 shows the percentage distribution of daily mean, maximum, and minimum temperatures for the three investigated areas. According to the percentage values of Figure 3, the maximum daily values of months May until September were observed at the urban and the suburban stations with values of 36–38 °C appearing more frequently (highest percentage), whereas at the rural station, values of 34–36 °C appeared more frequently (Figure 3). The minimum daily temperatures appeared slightly increased at the urban station with values between 22–24 °C, whereas at the other two stations, they were lower and fairly equal (Table 2) with values 20–22 °C (Figure 3). The mean daily temperatures of May until September ranged mainly between 28–30 °C for all stations, but a closer investigation showed a steady decrease of 0.5 °C during the thirty investigated years.

The mean monthly total precipitation was usually lower at the urban station during the months May, June, and September, whereas for the months July and August, due to the extremely low precipitation levels (Table 2), a significant variation between the three stations could not be corroborated.

Table 2. Average monthly maximum (Tmax), minimum (Tmin), and mean (Tmean) air temperatures (and daily absolute maximum and minimum ambient air temperatures) and total precipitation using data from years 1988–2017 for urban, suburban, and rural stations.

		May	June	July	August	September
Tmax (°C)	Urban	29.9 [17.5–41.1]	34.4 [17.7–44.9]	37.4 [25.4–44.8]	37.2 [27.9–44.5]	33.6 [23.9–41.9]
	Suburban	29.5 [18.1–41.5]	34.2 [23.7–45.4]	37.2 [29.7–43.6]	37.1 [30.2–46.2]	33.5 [24.4–41.1]
	Rural	29.1 [18.0–41.0]	33.4 [23.3–42.5]	36.2 [29.7–44.5]	36.1 [30.0–43.0]	33.0 [25–42.2]
Tmin (°C)	Urban	16.2 [9.2–23.2]	20.5 [12.3–29.1]	23.5 [17.0–29.4]	23.4 [18.8–29.7]	20.2 [13.8–27.0]
	Suburban	15.3 [7.4–23.7]	19.8 [10.2–28.7]	22.6 [15.5–30.2]	22.6 [16.3–30.1]	19.4 [13.0–26.9]
	Rural	15.2 [6.8–24.4]	19.5 [12.0–29.5]	22.5 [16.0–29.8]	22.6 [16.5–29.1]	19.6 [13.1–27.4]
Tmean (°C)	Urban	23.0 [14.5–31.0]	27.5 [16.6–35.6]	30.4 [24.1–36.1]	30.3 [24.5–36.2]	26.9 [19.8–34.0]
	Suburban	22.4 [14.3–31.0]	27.0 [17.0–35.7]	29.9 [23.0–36.3]	29.8 [24.4–38.2]	26.4 [19.5–33.6]
	Rural	22.1 [14.3–32.0]	26.4 [19.0–35.8]	29.4 [23.5–36.6]	29.3 [24.8–35.9]	26.3 [19.6–34.8]
Total Precipitation (mm)	Urban	18.9	7.0	3.7	2.4	4.5
	Suburban	24.8	13.8	4.6	1.6	11.5
	Rural	26.8	14.0	2.7	1.4	10.1

3.2. Temperature Anomalies

The term temperature anomaly (Tanomaly) means a deviation from a long-term average, with positive/negative Tanomaly values indicating that the observed temperature was warmer/cooler than the reference value. Reference values were computed on local scales over a defined time period, establishing a baseline from which the anomalies were calculated. This resulted in normalization of the data in order for them to be compared and combined to a more accurate temperature pattern with respect to normal climatic values of a specific region. The average maximum temperatures of each month (Tmax of Table 2) were considered as the baseline values from which anomalies were calculated and were used for the calculation of temperature anomalies.

3.3. Standardized Precipitation and Evapotranspiration Index (SPEI)

Droughts are identified by their effect at different levels, such as duration, intensity, magnitude, spatial extent, and onset, but there is not a physical variable to quantify them. Over the years, several drought indices have been developed with the most wide usage of the Palmer Drought Severity Index (PDSI) [31,32] and the Standardized Precipitation Index (SPI) [33,34]. PDSI is based on a simplified water balance equation that incorporates prior precipitation, moisture supply, runoff, and evaporation demand at the surface level [32], whereas SPI is based on precipitation anomalies and has the advantage of analyzing different temporal scales [33].

In this study, we utilized the Standardized Precipitation and Evapotranspiration Index (SPEI), which is a commonly used index that combines the sensitivity of PDSI with changes in evaporation demand and the multi-temporal nature of the SPI [35]. Several studies showed that SPEI more accurately captures the impacts of droughts on hydrological, agricultural, and ecological variables compared to SPI or PDSI. The SPEI allows comparison of drought severity through time and space since it can be calculated over a wide range of climates and is statistically robust with clear and comprehensible calculation procedure [35–37].

The following table (Table 3) shows the categorization of the area according to SPEI values. The SPEI allows the comparison of drought severity through time and intensity and can identify the onset and the end of drought episodes. For the calculation of SPEI, the preceding month's precipitation is required for the water balance equation. SPEI was calculated on a daily basis in order to relate drought episodes to soil water content and river discharge in headwater areas. Larger time scales are used to monitor drought conditions in different hydrological subsystems, such as reservoir and groundwater storages [38].

Table 3. Categorization according to the Standardized Precipitation-Evapotranspiration Index (SPEI) values.

SPEI Values	Categories
Over 2	Extreme Wet
1.5 to 2	Severe Wet
1 to 1.5	Moderate Wet
−1 to 1	Normal climate
−1.5 to −1	Moderate Dry
−2 to −1.5	Severe Dry
Less than −2	Extreme Dry

The SPEI index was calculated based on precipitation and potential evapotranspiration (PE), which was evaluated according to the SPEI package [36] in RStudio by implementing the Hargreaves equation and the log-logistic distribution of the water surplus or deficit. The Hargreaves equation [39] was preferred over other equations of potential evapotranspiration (Penman or Thornthwaite) due to its simplicity and accuracy, as it gives an estimate of the potential evapotranspiration based mainly on temperature adjusted for the sunshine hours per day and is given by:

$$PE = 0.0023 \cdot (Tmean + 17.8) \cdot (Tmax - Tmin)^{0.5} \cdot R_a \tag{1}$$

where Tmean, Tmax, and Tmin are mean, maximum, and minimum daily temperatures (Celsius), respectively, and Ra is the extra-terrestrial radiation ($MJm^{-2}day^{-1}$), which is calculated as:

$$R_\alpha = \frac{1440}{\pi} \cdot 0.082 \cdot (1 + 0.033 \cdot \cos\left(\frac{2\pi \cdot \text{Julian day}}{\text{Number of days in year (366 in leap year)}}\right)) \tag{2}$$

A simple measure of the water surplus or deficit for each analyzed day (Di) is then calculated as the difference between the precipitation (PR) and the PE of each day.

$$D_i = PR_i - PE_i \tag{3}$$

Vicente-Serrano et al. [35] further explored this water surplus or deficit at different time scales, adjusted it to a log-logistic probability distribution [F(D)], and proposed the climatic drought index SPEI [35].

According to Vicente-Serrano et al. [35], the standardized values of the log-logistic probability distribution [F(D)] and the soil water balance (W) values could be used for the SPEI calculation by following the classical approximation of Abramowitz and Stegun [40] and resulted in the following equation:

$$SPEI = W - \frac{C_0 + C_1 W + C_2 W^2}{1 + d_1 W + d_2 W^2 + d_3 W^3} \tag{4}$$

where the constants are C_0 = 2.515517, C_1 = 0.802853, C_2 = 0.010328, d_1 = 1.432788, d_2 = 0.189269, and d_3 = 0.001308. The average value of the SPEI is 0, and the standard deviation is 1. For this study, daily SPEI index was evaluated using the data of the investigated time period (years 1988–2017) for the three stations.

3.4. Retrospective Approach with Cross Correlation

Soil moisture is increased with precipitation, and this consequently modifies the total energy used by latent heat flux. Therefore, more energy is available for sensible heating, resulting in the increase of ambient air temperature [15]. The effects of precipitation may prolong for a number of days and may vary according to the investigated area. The lag period of the precipitation effect on lowering daily temperatures was found using the cross correlation function (CCF analysis), which computes the correlation between two variables, x and y. If ⊗ denotes correlation, then the cross-correlation function is defined as [41]:

$$R_{xy}(t) = x(t) \otimes y(t) = \int_{-\infty}^{\infty} x(h)\ y(t+h)\ dh \tag{5}$$

where y(t) are the precipitation levels shifted to the left by h-lag time, and x(t) is the temperature anomalies deviating from the average maximum temperatures of each month (Tmax of Table 2). The lag period used in our study was h = 0, 1, 2, ... , 30 days, thus we also had to employ daily precipitation levels in the month of April.

The cross-correlation analysis was also followed for correlating the temperature anomalies [x(t) component] with the SPEI [y(t) component]. The same lag period h = 0, 1, 2, ... , 30 days was used. The variances of the cross-correlation coefficient under the null hypothesis of zero correlation for both cross-correlation analyses were approximately 0.0002.

4. Results

4.1. Temperature Anomalies and Lag Period

In Section 3.1, the average maximum temperatures of each month were found (Tmax of Table 2) and were later used for the calculation of the temperature anomalies for the months May to September. These anomalies were divided into positive (above the average) or negative (below the average) values with the majority of them varying between −2 °C and 2 °C (as shown in Figure 4). Specifically, for months May to September and years 1988–2017 (inclusive), around 52.6%, 53.6%, and 56.7% of the temperature anomalies in the urban, the suburban, and the rural stations, respectively, varied between −2 °C and 2 °C. The most positive anomalies were found at the suburban station, and the most negative anomalies were found at the rural station. Moreover, about 2.6–3% of the temperature extremes exceeded the average monthly maximum temperature by 6 °C at all stations.

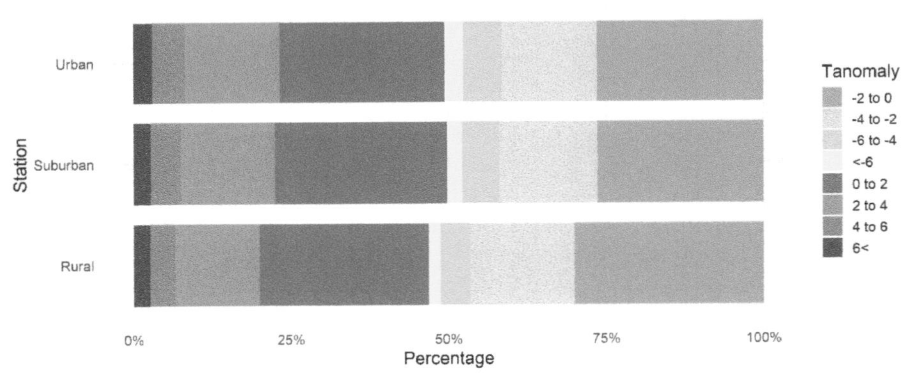

Figure 4. Air temperature anomalies above the average maximum temperatures of each month (Tmax of Table 2) for urban, suburban, and rural stations.

Figure 5 illustrates the results of the cross correlation analysis (Section 3.4) for the temperature anomalies happening after a precipitation event for the three investigated areas. The blue dashed lines in Figure 5 represent the significance limit at $\alpha = 0.05$ of $R_{xy}(t)$ (Equation (5)) in order to determine the statistical significance of a null-hypothesis. The variance of the cross-correlation coefficient under the null hypothesis of zero correlation for this study was approximately 0.0002, thus the approximate critical values (at the 5% level) were ±0.029 (to three decimal places). On a rainy day, day 0, an immediate drop of temperature appeared, which prolonged for six, seven, and nine days at the urban, the suburban, and the rural stations, respectively. The rural station was influenced by precipitation, resulting in a delayed increase of temperature after a rainfall event, which explained the higher percentage of negative anomalies as well as the relatively steady temperature profile without any extremities. Negative temperature anomalies prolonged for up to nine days at the rural station, signifying the importance of soil moisture for preventing extremely high temperatures in the summer. On the contrary, the urban station seemed the most susceptible to extremely high temperatures above the average maximum temperatures of each month (Tmax of Table 2), with the most days over 4 °C above the average maximum temperatures of each month. In urban areas, the urban environment resulted in high water runoff through the concrete structures and rapid evaporation of the overlay water, leading to a small decrease of temperature that only lasted for five to six days. He et al. [42] also indicated stronger impacts on diurnal temperature range extremes from short-term rather than long-term precipitation deficits and that low soil moisture due to precipitation deficits increase air temperatures through higher sensible heat flux [42].

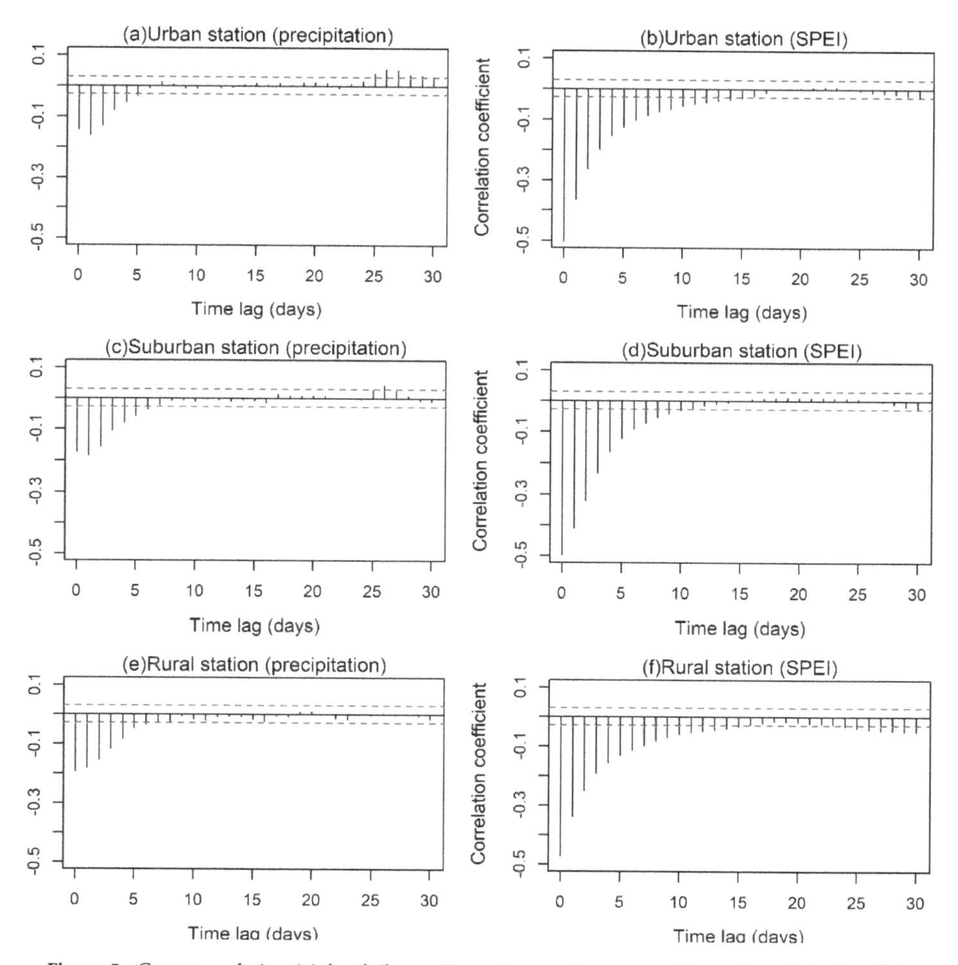

Figure 5. Cross correlation (y) for daily maximum temperature anomalies and precipitation (left column) or SPEI (right column) in the urban station, the suburban station, and the rural station. Dotted blue horizontal lines show 95% significance limits.

The cross-correlation analysis of SPEI with temperature anomalies revealed the stronger relationship and the importance of this index. The correlation coefficient increased from −0.2 for precipitation to −0.5 for SPEI, a percentage increase of 150%. Temperature anomalies and SPEI had a negative correlation and evolved concurrently, i.e., when one parameter increased, the other decreased, and vice versa. In the case of SPEI with regards to temperature anomalies, the lag period was significantly longer: 15, 11, and 16 days at the urban, the suburban, and the rural stations, respectively.

Figure 6 shows the time-series variation of daily temperature anomalies with respect to the daily accumulated precipitation. Negative temperature anomalies were clearly observed for a rainy day, as well as for the days following a precipitation event, suggesting local climatic variations strongly controlled by the evapotranspiration of small soil moisture after the precipitation event. From Figure 6, it was also noted that there was no significant trend towards increasing or decreasing temperature anomalies in Cyprus within the last 30 years during the summer period. In contrast, temperatures showed some changes over time, with average and minimum temperatures increasing, and this was accompanied by a significant decrease in the daily temperature range (DTR) [9].

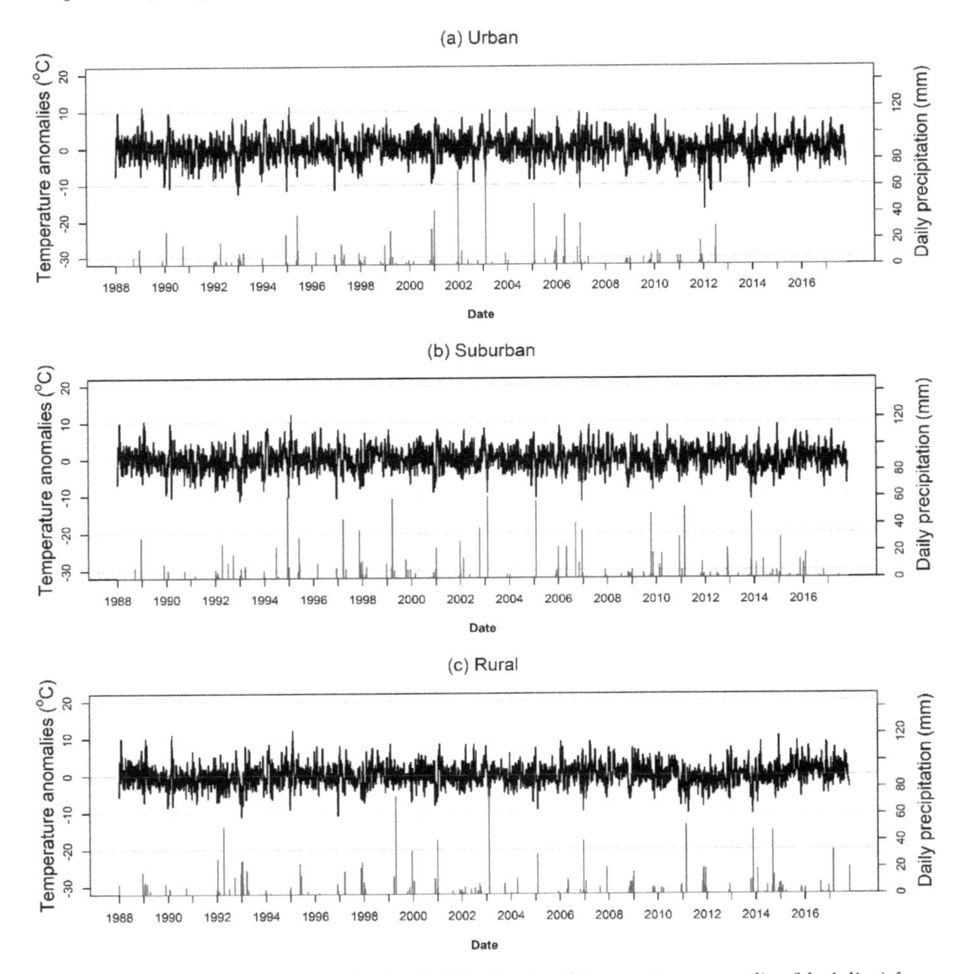

Figure 6. Time series of precipitation levels (blue bars) and temperature anomalies (black line) for months May to September for years 1988–2017.

Precipitation is directly linked to regional evapotranspiration, but this is trivial in cases of intense precipitation or areas under extreme drought. In the investigated areas of this study, moderate drought conditions dominated, thus it was expected that, after rainfall, the higher soil moisture would increase the evapotranspiration. Positive soil moisture anomaly led to a negative temperature anomaly mediated through a positive anomaly of evapotranspiration. Small soil moisture indicated a small evapotranspiration rate, which, according to Seneviratne et al. [15], is stronger in transitional zones between dry and wet climates. In the case of precipitation, the soil moisture may increase, leading to an evapotranspiration rate increase and a consequent decrease of temperature and negative temperature anomalies. Typically, low evapotranspiration rate is linked to lower energy used by latent heat flux and an increase in sensible heat flux and thus an increase in positive temperature anomalies. Therefore, even a small increase of the evapotranspiration rate after a precipitation event suggested higher energy used by latent heat rather than sensible heat flux, leading to fewer positive temperature anomalies compared to days not following a precipitation event.

4.2. Analysis of SPEI

The SPEI was calculated for years 1988–2017 for the months of May to September based on the temperatures and the precipitation of the preceding days. The regression lines of Table 4 show the existence of a negative relationship between the parameters SPEI and Tanomalies. They could not be used for the estimation of the daily variation of Tanomalies, as there was large variation around the mean values and the respective standard deviations of the two parameters (SPEI and Tanomalies). A statistically significant increasing trend for the time-series of the mean values of mean and minimum ambient air temperatures (Table 1) was previously proven. However, the regression lines of Table 4 reveal the higher decreasing trend of SPEI during the thirty investigated years at the suburban and the rural station, which was probably attributed to by external factors (change of land cover, meteorological conditions, etc.). More frequent temperature anomalies were observed in August for the urban and the suburban stations and in July for the rural station.

Table 4. Monthly mean and standard deviation (sd) values of SPEI and temperature anomalies ($T_{anomaly}$) for urban, suburban, and rural station.

Month		Urban	Suburban	Rural
May	$T_{anomaly}$ (mean ± sd)	−0.002 ± 4.20	0.046 ± 4.01	−0.007 ± 3.82
	SPEI (mean ± sd)	0.25 ± 0.74	0.31 ± 0.69	0.30 ± 0.72
June	$T_{anomaly}$ (mean ± sd)	0.011 ± 3.27	−0.048 ± 3.19	−0.031 ± 3.04
	SPEI (mean ± sd)	−0.39 ± 0.58	−0.37 ± 0.57	−0.37 ± 0.59
July	$T_{anomaly}$ (mean ± sd)	−0.045 ± 2.40	−0.034 ± 2.20	0.045 ± 2.17
	SPEI (mean ± sd)	−0.913 ± 0.54	−0.943 ± 0.46	−0.925 ± 0.51
August	$T_{anomaly}$ (mean ± sd)	0.040 ± 2.08	0.025 ± 2.02	−0.011 ± 1.93
	SPEI (mean ± sd)	−0.93 ± 0.43	−0.989 ± 0.42	−0.931 ± 0.45
September	$T_{anomaly}$ (mean ± sd)	0.001 ± 2.83	0.021 ± 2.76	0.027 ± 2.49
	SPEI (mean ± sd)	−0.316 ± 0.55	−0.309 ± 0.58	−0.358 ± 0.54

According to Figure 7, the urban, the suburban, and the rural stations were mainly characterized by a normal climate (SPEI between −1 to 1) with 73.5%, 73.9%, and 74.2% SPEI values, respectively, for the five months. The highest percentages (77.6 to 86.9%) of normal climatic conditions (SPEI between −1 and 1) were observed in May, June, and September.

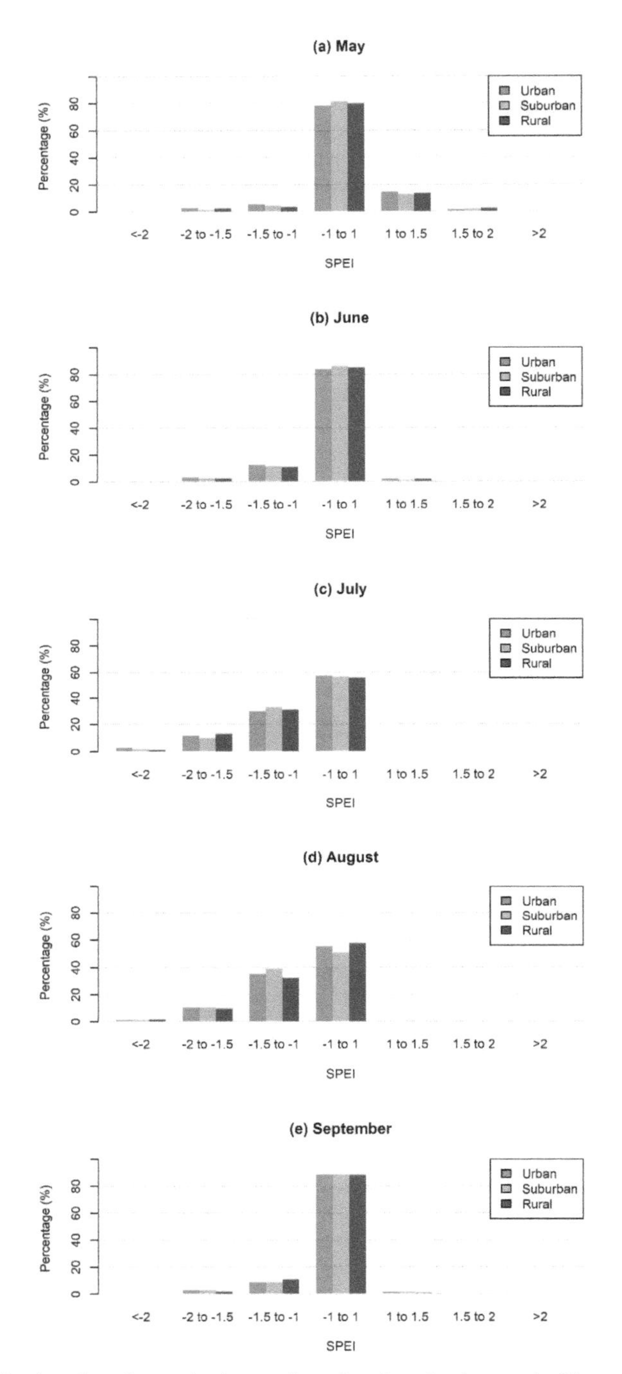

Figure 7. SPEI values for urban, suburban, and rural stations for the months May to September of years 1988–2017.

As depicted in Figure 7, July and August were the major drought months in the study area with SPEI below −1, contributing to about 35% of the total SPEI values. It is worth noting that, during July and August, no days were observed with wet conditions (SPEI over 1), whereas in May, a small occurrence of wet conditions (SPEI over 1) associated with negative Tanomalies was observed at a percentage of 13.4–14.5% (Figure 7 and Table 5).

Table 5. Percentage of occurrence of $T_{anomalies}$ for each station under wet conditions (SPEI > 1), dry conditions (SPEI < −1), and normal climatic conditions (−1 < SPEI < 1).

				May				
	Urban			Suburban			Rural	
	SPEI < −1	SPEI > 1		SPEI < −1	SPEI > 1		SPEI < −1	SPEI > 1
$T_{anomaly} > 0$	7.3%	0.5%	$T_{anomaly} > 0$	5.3%	0.4%	$T_{anomaly} > 0$	5.2%	0.6%
$T_{anomaly} < 0$	0.2%	14.3%	$T_{anomaly} < 0$	0.1%	13.4%	$T_{anomaly} < 0$	0.1%	14.5%
all $T_{anomaly}$	−1 < SPEI < 1	77.6%	all $T_{anomaly}$	−1 < SPEI < 1	80.8%	all $T_{anomaly}$	−1 < SPEI < 1	79.6%
				June				
	Urban			Suburban			Rural	
	SPEI < −1	SPEI > 1		SPEI < −1	SPEI > 1		SPEI < −1	SPEI > 1
$T_{anomaly} > 0$	12.6%	0%	$T_{anomaly} > 0$	11.2%	0%	$T_{anomaly} > 0$	11.2%	0.1%
$T_{anomaly} < 0$	2.3%	1.4%	$T_{anomaly} < 0$	2.1%	1.0%	$T_{anomaly} < 0$	2.2%	1.8%
all $T_{anomaly}$	−1 < SPEI < 1	83.8	all $T_{anomaly}$	−1 < SPEI < 1	85.7%	all $T_{anomaly}$	−1 < SPEI < 1	84.7%
				July				
	Urban			Suburban			Rural	
	SPEI < −1	SPEI > 1		SPEI < −1	SPEI > 1		SPEI < −1	SPEI > 1
$T_{anomaly} > 0$	29.5%	0%	$T_{anomaly} > 0$	29.7%	0%	$T_{anomaly} > 0$	28.7%	0%
$T_{anomaly} < 0$	12.4%	0%	$T_{anomaly} < 0$	13.3%	0%	$T_{anomaly} < 0$	15.5%	0%
all $T_{anomaly}$	−1 < SPEI < 1	58.2%	all $T_{anomaly}$	−1 < SPEI < 1	57.0%	all $T_{anomaly}$	−1 < SPEI < 1	55.8%
				August				
	Urban			Suburban			Rural	
	SPEI < −1	SPEI > 1		SPEI < −1	SPEI > 1		SPEI < −1	SPEI > 1
$T_{anomaly} > 0$	32.0%	0%	$T_{anomaly} > 0$	33.7%	0%	$T_{anomaly} > 0$	27.4%	0%
$T_{anomaly} < 0$	12.3%	0%	$T_{anomaly} < 0$	15.3%	0%	$T_{anomaly} < 0$	15.2%	0%
all $T_{anomaly}$	−1 < SPEI < 1	55.7%	all $T_{anomaly}$	−1 < SPEI < 1	51.1%	all $T_{anomaly}$	−1 < SPEI < 1	57.4%
				September				
	Urban			Suburban			Rural	
	SPEI < −1	SPEI > 1		SPEI < −1	SPEI > 1		SPEI < −1	SPEI > 1
$T_{anomaly} > 0$	10.1%	0%	$T_{anomaly} > 0$	10.1%	0%	$T_{anomaly} > 0$	10.3%	0%
$T_{anomaly} < 0$	1.0%	1.0%	$T_{anomaly} < 0$	0.7%	1.2%	$T_{anomaly} < 0$	1.1%	0.3%
all $T_{anomaly}$	−1 < SPEI < 1	87.9%	all $T_{anomaly}$	−1 < SPEI < 1	88.0%	all $T_{anomaly}$	−1 < SPEI < 1	88.2%

Dry conditions with SPEI lower than −1 were associated with positive temperature anomalies (Tanomalies > 0 °C) at percentages from 10.7 to 31.7%. Dry conditions were associated with negative temperature anomalies (Tanomalies < 0 °C) at percentages from 1.4 to 15.4% (Table 5). This frequency was increased in July and August, confirming the overall drought in the area during these two summer months.

In summary, there were no large discrepancies in the monthly SPEI values between the three areas, but more severe and extreme dry conditions (SPEI less than −1.5) occurred at the rural area in July and August.

To quantitatively describe the SPEI, we calculated the percentage of positive and negative temperature anomalies for SPEI lower than −1 (drought conditions) and higher than 1 (wet conditions). The results are shown in Table 5. In May, the percentage of negative temperature anomalies with SPEI > 1 was greater than the percentage of positive anomalies combined with either SPEI < −1 or SPEI > 1, indicating a greater proportion of low temperatures occurred under wet conditions. For the months June to September, the percentage of positive temperature anomalies with SPEI < −1 was greater than the percentage of negative anomalies. Zero positive temperature anomalies were found for SPEI > 1 for months June to September, which indicated that all higher air temperatures occurred during dry conditions. No wet climatic conditions appeared during the summer, mainly due to the lack of precipitation. Positive temperature anomalies reached a peak in August under dry conditions, with occurrences of 32.0%, 33.7%, and 27.4% at the urban, the suburban, and the rural stations, respectively. Most of the temperature anomalies occurred for SPEI values between −1 and 1, with greater values in June and September. Comparison of the percentage values between the three stations revealed that most positive temperature anomalies occurred in the urban and the suburban areas, and most negative temperature anomalies occurred in the rural area.

4.3. Concurrent Drought and Hot Days

In the next stage, we investigated the occurrence of positive temperature anomalies above the average maximum temperatures of each month (Tmax of Table 2) with respect to the SPEI in order to assess whether they appeared more frequently under dry conditions. The regression analysis showed the monthly relation between the two variables—SPEI and temperature anomalies (T anomalies). The results according to Figure 8 and Tables 6 and 7 showed:

- R^2 values (Table 6) for SPEI and Tanomaly showed that there was an overall significant linear relationship between the two parameters, which varied from 0.2 to 0.57 (for more than 900 degrees of freedom) for each investigated area. These relatively low R^2 values are not uncommon in large datasets because the significance of the slope is due to the number of elements in the dataset.
- Under normal climatic conditions (SPEI varying between −1 to 1) independent from the values of Tanomalies, we observed, for all the months, that the frequency of the pair SPEI/Tanomalies was generally the same at all stations (72.6% for urban, 72.5% for suburban, 73.1% for rural), representing a uniform climatic behavior in the wider range of Nicosia (Table 5).
- Figure 8 and Table 6 portray the monthly temperature anomalies and their trends for different SPEI. Using a linear regression model, the rate of change was defined by the slope of the regression line and differed in each investigated area and month. The linear regression lines for months July and August were almost identical at all three stations with slopes −2.34 (July) and −2.51 (August) for the urban, −2.46 (July) and −2.56 (August) for the suburban, and −2.87 (July) and −3.03 (August) for the rural station. The linear regression line's slope for May varied significantly with values −4.13, −4.42, and −3.69 for the urban, the suburban, and the rural stations, respectively, with higher occurrence of severe to moderate wet conditions (SPEI > 1) that were associated with positive as well as negative Tanomalies (Table 6).
- The negative slope of the linear regression lines was larger in May (varying from −3.69 to −4.14) and smaller in July and August (varying from −2.01 to −2.56) at all three stations, confirming the larger effects of evapotranspiration and precipitation that existed during the months March to April on the values of Tanomalies for the month of May.
- With the use of t-test analysis, the absolute values of $|t_A|$ and $|t_B|$ were calculated in order to check the statistical significance of A and B coefficients of the linear regression lines with:

$$T_{anomaly} = A + B \cdot SPEI \tag{6}$$

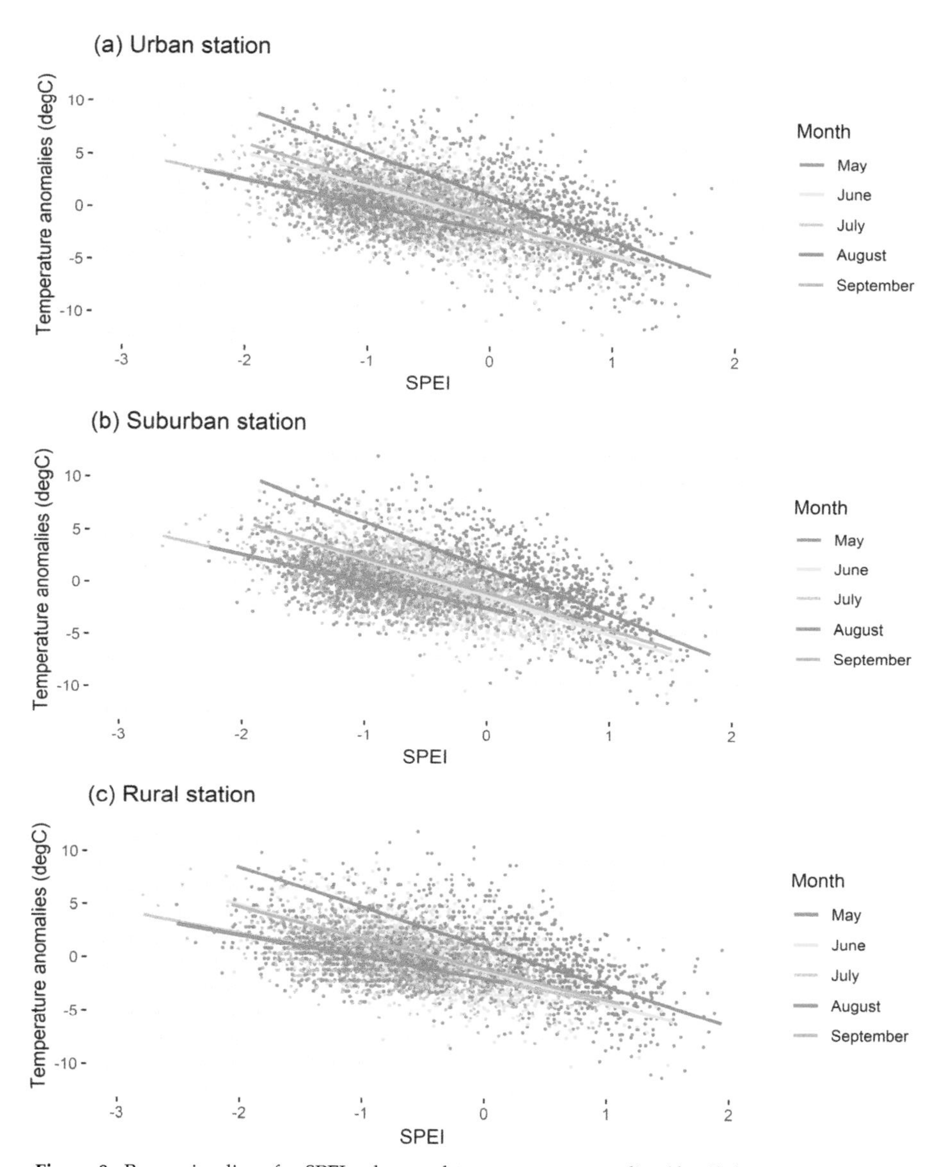

Figure 8. Regression lines for SPEI values and temperature anomalies (degC) for months May to September of years 1988–2017 for (**a**) urban station, (**b**) suburban station, and (**c**) rural station.

Table 6. Linear regression analysis of temperature anomalies with respect to SPEI for the three stations and for the months May to September of the years 1988–2017, showing the regression equation ($T_{anomaly} = A + B \cdot (SPEI)$), the t-test of the statistical significance of A and B, and the adjusted R^2 correlation coefficients.

Month	Station	A	B	t_A	t_B	R^2
May	Urban	1.025	−4.127	**10.17**	**−31.70**	**0.52**
	Suburban	1.403	−4.417	**14.86**	**−35.12**	**0.57**
	Rural	1.108	−3.687	**11.28**	**−29.14**	**0.47**
June	Urban	−1.236	−3.220	**−11.35**	**−21.09**	**0.33**
	Suburban	−1.338	−3.535	**−13.81**	**−24.61**	**0.40**
	Rural	−1.139	−3.027	**−11.79**	**−21.77**	**0.35**
July	Urban	−2.338	−2.512	**−18.27**	**−20.80**	**0.32**
	Suburban	−2.572	−2.461	**−17.86**	**−19.62**	**0.29**
	Rural	−1.921	−2.127	**−15.17**	**−17.76**	**0.25**
August	Urban	−2.152	−2.339	**−15.17**	**−17.03**	**0.24**
	Suburban	−2.510	−2.563	**−17.28**	**−18.94**	**0.28**
	Rural	−1.878	−2.006	**−14.65**	**−16.19**	**0.22**
September	Urban	−1.128	−3.577	**−14.45**	**−29.03**	**0.48**
	Suburban	−1.025	−3.395	**−13.92**	**−30.15**	**0.50**
	Rural	−0.997	−2.865	**−12.88**	**−24.06**	**0.39**

Table 7. Paired t-test for the statistical significance between $A_i.A_j$ and $B_i.B_j$ (for t values greater than 1.96, the differences were statistically significant for $\alpha = 0.05$).

	A			B		
	Urban/Suburban	Urban/Rural	Suburban/Rural	Urban/Suburban	Urban/Rural	Suburban/Rural
May	1.057	1.160	0.189	1.598	**−2.420**	**−4.087**
June	0.833	0.721	−0.096	1.518	−0.935	**−2.558**
July	−0.821	−1.111	−0.361	0.337	**−2.260**	**−2.506**
August	−1.186	0.696	1.840	1.161	−1.801	**−3.035**
September	0.012	−1.294	−1.282	−1.092	**−4.156**	**−3.236**

According to Table 6, for all cases, the coefficients A and B were found to be statistically significant (bold values) with $|t_A| > t_{0.05} = 1.96$ and $|t_B| > t_{0.05} = 1.96$.

- Paired samples t-test was employed to compare the mean difference in coefficients A and B between the different pairs of stations. The results are presented in Table 7 for the pairs urban/suburban, urban/rural, and suburban/rural for all months. Statistically significant mean difference was obtained ($t > 1.96$; $\alpha < 0.05$) at 95% level of significance. The results showed that the coefficient A was considered statistically equal for all pairs, indicating that the three investigated areas were nearby. The coefficient B defined the slope of the linear regression lines, which was a different trend of variation between the time series and suggested that external factors (land cover, meteorological conditions, etc.) differently affected the three stations during the thirty investigated years. The coefficient B was considered statistically equal ($t_{i,j} < 1{,}96$) in all the investigated months for the pair urban/suburban. For the pair urban/rural for months May, July, and September, the t-test was considered statistically significant. For the pair suburban/rural, the t-test showed that the B coefficients were statistically significant for all the months, which was attributed to a faster development of construction in the suburban area in relation to the urban area when both were compared to the rural area.
- The regression lines that were determined (Table 6 and Figure 8) only showed the existence of a negative relationship between the parameters SPEI and Tanomaly, but they could not be used for the estimation of the daily value of Tanomaly.

5. Discussion and Conclusions

Through linear and cross-correlation statistical analysis, this study examined the compound effect of precipitation levels and evapotranspiration rates of the preceding days to summer temperature anomalies for years 1988–2017. The observations of the time-series figure (Figure 6) and the cross-correlation results showed that the cooling effect of precipitation was higher and lasted more in rural and suburban areas compared to urban areas, a fact directly related to the evaporation potential of the area concerned. We showed that precipitation was the dominant driving force of positive temperature anomalies and that varying evapotranspiration rates contributed to the development of moderate to severe drought in the investigated areas.

Particularly, the investigation of temperature anomalies showed a higher correlation for the concurrent month's precipitation compared with precipitation in the preceding months, suggesting that moisture was depleted faster. This showed that there was a lag effect of soil moisture memory of six, six, and nine days in the urban, the suburban, and the rural areas, respectively. In warmer areas (urban and suburban areas), the larger evaporative demand from the atmosphere exacerbated the existing drought conditions and its impacts. Also, the higher urban and suburban temperatures (Table 2) compared to the rural area could significantly reduce the natural storage of water. With view of the precipitation events, the negative temperature anomalies suggested local climatic variations strongly controlled by the evapotranspiration of small soil moisture after the precipitation event. The SPEI was later used that employed both precipitation and evapotranspiration rates to characterize dry or wet conditions. The cross-correlation analysis of SPEI with temperature anomalies revealed the stronger relationship with negative correlation coefficient of −0.5 and highlighted the importance of this index. In the case of SPEI with regards to temperature anomalies, the lag periods according to the cross-correlation analysis were significantly longer: 15, 11, and 16 days at the urban, the suburban, and the rural stations, respectively. The higher surface albedo of the urban infrastructure may have led to additional warming. This does not necessarily translate to drier conditions and longer droughts, but it creates challenges for better water reservoir management.

According to this study, the SPEI has a high correlation with temperature anomalies and may be considered as a key tool for the identification of abnormal weather conditions and extremely high temperatures. Moreover, it confirmed that rainfall events combined with evapotranspiration, which could be effectively represented by SPEI index variation, may be the main regulators of soil moisture rather than the amount of monthly rainfall [43,44]. In the results section, the temperature anomalies were inversely correlated with precipitation anomalies, and the SPEI index and the linear regression coefficients were found. High temperatures during the summer months may be understood by the investigation of the soil moisture to understand the impact of soil storage memory on ambient air temperatures. Further analysis could focus on the division of temperature anomalies based on the amount of rainfall as well as the intervals between rainfall events. We should consider the effects of not only precipitation but also evapotranspiration in future studies to better understand the length of extreme weather conditions.

Further analysis focused on the statistical investigation of the linear regression lines of the SPEI with temperature anomalies for the three stations and for each month. The results of the paired t-test for the statistical significance showed that the coefficients A of Table 7 were considered statistically equal between them for all pairs, indicating that the three investigated areas were nearby. The B coefficients suggested that external factors (land cover, meteorological conditions, etc.) differently affected the three stations during the thirty investigated years. This study focused on the analysis of the effect of precipitation during the summer period on temperatures and particularly the deviation of temperature from the mean monthly value. The spatial investigation revealed a similar climatic profile in all three investigated areas but showed a noteworthy different lag effect of precipitation. Particularly, precipitation in rural areas led to a longer decrease of temperature compared to the urban and the suburban areas because the wet ground favored the increased evapotranspiration and the decrease of sensible heat flux. Later, the investigation of SPEI further supported the above statement, because SPEI was strongly negatively correlated with positive temperature anomalies.

Future work should focus on the effect of the intervals between precipitation events in urban, suburban, and rural areas. In this study, the semi-arid climate in Cyprus and the infrequent precipitation allowed a more comprehensive understanding of the lag effect of precipitation during the dry period (summer) in areas with different land cover. The lag period may vary seasonally; therefore, further investigation during the winter is necessary. The investigation of the transitional phase of dry and wet climates in Cyprus will likely confirm the strong soil-moisture climate coupling, which is the strong dependency of evapotranspiration on soil moisture during the dry periods and the little impact of soil moisture on evapotranspiration during the wet periods.

Author Contributions: M.S. conceived the research topic. A.P. obtained the datasets, created the figures and analyzed the results. I.L. designed the methodology and did the statistical analysis of the data. All authors (A.P., M.S., I.L. and C.C.) contributed in the discussion of the results and reviewed the manuscript.

Funding: This research received no external funding.

Acknowledgments: The authors are grateful to the Ministry of Agriculture, Rural Development and Environment (MADRE) of the Republic of Cyprus for the Department of Meteorology historical meteorological data. Special thanks to Marinos Eliades for the creation of Figure 1 in ArcGis software version 10.3 (www.ESRI.com).

Conflicts of Interest: The authors declare no competing interests.

References

1. Yiou, P.; Vautard, R.; Naveau, P.; Cassou, C. Inconsistency between atmospheric dynamics and temperatures during the exceptional 2006/2007 fall/winter and recent warming in Europe. *Geophys. Res. Lett.* **2007**, *34*, 1–7. [CrossRef]
2. Fischer, E.M.; Seneviratne, S.I.; Vidale, P.L.; Lüthi, D.; Schär, C. Soil moisture-atmosphere interactions during the 2003 European summer heat wave. *J. Clim.* **2007**, *20*, 5081–5099. [CrossRef]
3. Miralles, D.G.; Gentine, P.; Seneviratne, S.I.; Teuling, A.J. Land–atmospheric feedbacks during droughts and heatwaves: State of the science and current challenges. *Ann. N. Y. Acad. Sci.* **2019**, *1436*, 19–35. [CrossRef] [PubMed]
4. Cassou, C.; Terray, L.; Hurrell, J.W.; Deser, C. North Atlantic winter climate regimes: Spatial asymmetry, stationarity with time, and oceanic forcing. *J. Clim.* **2004**. [CrossRef]
5. Feudale, L.; Shukla, J. Role of Mediterranean SST in enhancing the European heat wave of summer 2003. *Geophys. Res. Lett.* **2007**. [CrossRef]
6. Doblas-Reyes, F.J.; García-Serrano, J.; Lienert, F.; Biescas, A.P.; Rodrigues, L.R.L. Seasonal climate predictability and forecasting: Status and prospects. *Wiley Interdiscip. Rev. Clim. Chang.* **2013**, *4*, 245–268. [CrossRef]
7. Trenberth, K.E.; Shea, D.J. Relationships between precipitation and surface temperature. *Geophys. Res. Lett.* **2005**. [CrossRef]
8. LeMone, M.A.; Grossman, R.L.; Chen, F.; Ikeda, K.; Yates, D. Choosing the Averaging Interval for Comparison of Observed and Modeled Fluxes along Aircraft Transects over a Heterogeneous Surface. *J. Hydrometeorol.* **2003**. [CrossRef]
9. Pyrgou, A.; Santamouris, M.; Livada, I. Spatiotemporal Analysis of Diurnal Temperature Range: Effect of Urbanization, Cloud Cover, Solar Radiation, and Precipitation. *Climate* **2019**, *7*, 89. [CrossRef]
10. Hirschi, M.; Seneviratne, S.I.; Alexandrov, V.; Boberg, F.; Boroneant, C.; Christensen, O.B.; Formayer, H.; Orlowsky, B.; Stepanek, P. Observational evidence for soil-moisture impact on hot extremes in southeastern Europe. *Nat. Geosci.* **2011**, *4*, 17–21. [CrossRef]
11. Vautard, R.; Yiou, P.; D'Andrea, F.; de Noblet, N.; Viovy, N.; Cassou, C.; Polcher, J.; Ciais, P.; Kageyama, M.; Fan, Y. Summertime European heat and drought waves induced by wintertime Mediterranean rainfall deficit. *Geophys. Res. Lett.* **2007**, *34*, 1–5. [CrossRef]
12. Seneviratne, S.I.; Lüthi, D.; Litschi, M.; Schär, C. Land-atmosphere coupling and climate change in Europe. *Nature* **2006**, *443*, 205–209. [CrossRef] [PubMed]
13. Vidale, P.L.; Lüthi, D.; Wegmann, R.; Schär, C. European summer climate variability in a heterogeneous multi-model ensemble. *Clim. Chang.* **2007**, *81*, 209–232. [CrossRef]

14. Rowell, D.P.; Jones, R.G. Causes and uncertainty of future summer drying over Europe. *Clim. Dyn.* **2006**, *27*, 281–299. [CrossRef]
15. Seneviratne, S.I.; Corti, T.; Davin, E.L.; Hirschi, M.; Jaeger, E.B.; Lehner, I.; Orlowsky, B.; Teuling, A.J. Investigating soil moisture-climate interactions in a changing climate: A review. *Earth-Sci. Rev.* **2010**, *99*, 125–161. [CrossRef]
16. McHugh, T.A.; Morrissey, E.M.; Reed, S.C.; Hungate, B.A.; Schwartz, E. Water from air: An overlooked source of moisture in arid and semiarid regions. *Sci. Rep.* **2015**, *5*, 13767. [CrossRef] [PubMed]
17. Wang, B.; Zha, T.S.; Jia, X.; Wu, B.; Zhang, Y.Q.; Qin, S.G. Soil moisture modifies the response of soil respiration to temperature in a desert shrub ecosystem. *Biogeosciences* **2014**. [CrossRef]
18. Agam, N.; Berliner, P.R. Dew formation and water vapor adsorption in semi-arid environments—A review. *J. Arid Environ.* **2006**, *65*, 572–590. [CrossRef]
19. Liu, D.; Wang, G.; Mei, R.; Yu, Z.; Yu, M. Impact of initial soil moisture anomalies on climate mean and extremes over Asia. *J. Geophys. Res.* **2014**, *119*, 529–545. [CrossRef]
20. Naumann, G.; Alfieri, L.; Wyser, K.; Mentaschi, L.; Betts, R.A.; Carrao, H.; Spinoni, J.; Vogt, J.; Feyen, L. Global Changes in Drought Conditions Under Different Levels of Warming. *Geophys. Res. Lett.* **2018**, *45*, 3285–3296. [CrossRef]
21. Eliades, M.; Bruggeman, A.; Djuma, H.; Lubczynski, M. Tree Water Dynamics in a Semi-Arid, Pinus brutia Forest. *Water* **2018**, *10*, 1039. [CrossRef]
22. Feller, U.; Vaseva, I.I. Extreme climatic events: Impacts of drought and high temperature on physiological processes in agronomically important plants. *Front. Environ. Sci.* **2014**, *2*, 39. [CrossRef]
23. Pyrgou, A.; Santamouris, M. Increasing Probability of Heat-Related Mortality in a Mediterranean City Due to Urban Warming. *Int. J. Environ. Res. Public Health* **2018**, *15*, 1571. [CrossRef] [PubMed]
24. Peel, M.C.; Finlayson, B.L.; McMahon, T.A. Updated world map of the Koppen-Geiger climate classification. *Hydrol. Earth Syst. Sci.* **2007**, *11*, 1633–1644. [CrossRef]
25. Υπουργείο Εσωτερικών, Τ. Π. και Ο. Ισχύοντα Δημοσιευμένα Σχέδια Ανάπτυξης. Available online: http://www.moi.gov.cy/moi/tph/tph.nsf/page72_gr/page72_gr?OpenForm (accessed on 29 July 2019).
26. Cyprus, R. of Department of Meteorology, Cyprus. Available online: http://www.moa.gov.cy/moa/ms/ms.nsf/DMLannual_en/DMLannual_en?OpenDocument (accessed on 25 April 2019).
27. Begert, M.; Schlegel, T.; Kirchhofer, W. Homogeneous temperature and precipitation series of Switzerland from 1864 to 2000. *Int. J. Climatol.* **2005**, *25*, 65–80. [CrossRef]
28. Klein Tank, A.M.G.; Wijngaard, J.B.; Können, G.P.; Böhm, R.; Demarée, G.; Gocheva, A.; Mileta, M.; Pashiardis, S.; Hejkrlik, L.; Kern-Hansen, C.; et al. Daily dataset of 20th-century surface air temperature and precipitation series for the European Climate Assessment. *Int. J. Climatol.* **2002**, *22*, 1441–1453. [CrossRef]
29. Eischeid, J.K.; Pasteris, P.A.; Diaz, H.F.; Plantico, M.S.; Lott, N.J. Creating a Serially Complete, National Daily Time Series of Temperature and Precipitation for the Western United States JON. *J. Appl. Meteorol.* **2000**, *39*, 1580–1591. [CrossRef]
30. Klein, T.; Konnen, G. Trends in Indices of Daily Temperature and Precipitation Extremes in Europe, 1946–99. *J. Clim.* **2003**, *16*, 3665–3680. [CrossRef]
31. Wells, N.; Goddard, S.; Hayes, M.J. A self-calibrating Palmer Drought Severity Index. *J. Clim.* **2004**. [CrossRef]
32. Alley, W.M. The Palmer Drought Severity Index: Limitations and Assumptions. *J. Clim. Appl. Meteorol.* **2002**. [CrossRef]
33. Guttman, N.B. Accepting the standardized precipitation index: A calculation algorithm. *J. Am. Water Resour. Assoc.* **1999**. [CrossRef]
34. Hayes, M.J.; Svoboda, M.D.; Wilhite, D.A.; Vanyarkho, O.V. Monitoring the 1996 Drought Using the Standardized Precipitation Index. *Bull. Am. Meteorol. Soc.* **1999**. [CrossRef]
35. Vicente-Serrano, S.M.; Beguería, S.; López-Moreno, J.I. A multiscalar drought index sensitive to global warming: The standardized precipitation evapotranspiration index. *J. Clim.* **2010**, *23*, 1696–1718. [CrossRef]
36. Begueria, S.; Vicente-Serrano, S.M. Calculation of the Standardized Precipitation-Evapotranspiration Index. *Packag. SPEI* **2003**. [CrossRef]
37. Stagge, J.; Tallaksen, L. Standardized precipitation-evapotranspiration index (SPEI): Sensitivity to potential evapotranspiration model and parameters. *Int. Assoc. Hydrol. Sci.* **2014**, *10*, 367–373.
38. Santiago, B.; Borja Latorre, F.; Vicente-Serrano, R.S. About SPEI. Available online: https://spei.csic.es/home.html (accessed on 18 April 2019).

39. Thornthwaite, C.W. An Approach toward a Rational Classification of Climate. *Geogr. Rev.* **1948**. [CrossRef]
40. Abramowitz, M.; Stegun, I.A.; Miller, D. Handbook of Mathematical Functions With Formulas, Graphs and Mathematical Tables (National Bureau of Standards Applied Mathematics Series No. 55). *J. Appl. Mech.* **1965**. [CrossRef]
41. Sadiku, M.N.O.; Musa, S.M.; Nelatury, S.R. Correlation: A Brief Introduction. *Int. J. Electr. Eng. Educ.* **2014**, *51*, 93–99. [CrossRef]
42. He, B.; Huang, L.; Wang, Q. Precipitation deficits increase high diurnal temperature range extremes. *Sci. Rep.* **2015**, *5*, 12004. [CrossRef]
43. Laporte, M.F.; Duchesne, L.C.; Wetzel, S. Effect of rainfall patterns on soil surface CO2 efflux, soil moisture, soil temperature and plant growth in a grassland ecosystem of northern Ontario, Canada: Implications for climate change. *BMC Ecol.* **2002**, *2*, 10. [CrossRef]
44. Fay, P.A.; Carlisle, J.D.; Danner, B.T.; Lett, M.S.; McCarron, J.K.; Stewart, C.; Knapp, A.K.; Blair, J.M.; Collins, S.L. Altered Rainfall Patterns, Gas Exchange, and Growth in Grasses and Forbs. *Int. J. Plant Sci.* **2002**, *163*, 549–557. [CrossRef]

MDPI
St. Alban-Anlage 66
4052 Basel
Switzerland
Tel. +41 61 683 77 34
Fax +41 61 302 89 18
www.mdpi.com

Climate Editorial Office
E-mail: climate@mdpi.com
www.mdpi.com/journal/climate

Lightning Source UK Ltd.
Milton Keynes UK
UKHW052139120922
408735UK00002B/119